Issues in Open Research Data

Edited by
Samuel A. Moore

]u[

ubiquity press
London

Published by
Ubiquity Press Ltd.
Gordon House
29 Gordon Square
London WC1H 0PP
www.ubiquitypress.com

Text © The Authors 2014

First published 2014

Cover Illustration by Antonio Roberts
Pure Data File Killer - Bliss (sgi)
CC BY-SA
https://flic.kr/p/h4RV3y

ISBN (Paperback): 978-1-909188-30-3
ISBN (PDF): 978-1-909188-32-7
ISBN (EPub): 978-1-909188-31-0
ISBN (Kindle): 978-1-909188-41-9

DOI: http://dx.doi.org/10.5334/ban

This work is licensed under the Creative Commons Attribution 3.0 Unported License. To view a copy of this license, visit http://creativecommons.org/licenses/by/3.0/ or send a letter to Creative Commons, 444 Castro Street, Suite 900, Mountain View, California, 94041, USA. This license allows for copying any part of the work for personal and commercial use, providing author attribution is clearly stated.

Chapters 2, 4 and 5 have been previously published elsewhere and have been reproduced here under their CC-BY license, as well as the authors agreement.

Suggested citation:
Moore, S. A. (ed.) 2014 *Issues in Open Research Data*. London: Ubiquity Press. DOI: http://dx.doi.org/10.5334/ban

To read the free, open access version of this book online, visit http://dx.doi.org/10.5334/ban or scan this QR code with your mobile device:

Contents

Acknowledgements	v
Editor's Introduction (Samuel A. Moore)	1
Open Content Mining (Peter Murray-Rust, Jennifer C. Molloy and Diane Cabell)	11
The Need to Humanize Open Science (Eric C. Kansa)	31
Data Sharing in a Humanitarian Organization: The Experience of Médecins Sans Frontières (Unni Karunakara)	59
Why Open Drug Discovery Needs Four Simple Rules for Licensing Data and Models (Antony J. Williams, John Wilbanks and Sean Ekins)	77
Open Data in the Earth and Climate Sciences (Sarah Callaghan)	89
Open Minded Psychology (Wouter van den Bos, Mirjam A. Jenny and Dirk U. Wulff)	107
Open Data in Health Care (Tom Pollard and Leo Anthony Celi)	129
Open Research Data in Economics (Velichka Dimitrova)	141
Open Data and Palaeontology (Ross Mounce)	151

Acknowledgements

The publication of this book was made possible by Bibliophiles, Benefactors and Supporters via unglue.it. The following generous people paid for the book processing charges:

Bibliophiles

Agile Geoscience: *Free as in Science!*
Christian Jones: *Dedicated to those who can't afford to give, can't afford to share, can't afford to read; we're all in this together.*

Benefactors

Andy Byers
margaretehenderson
Pietro G-L - @p_gl
Todd Vision
wilbanks

Supporters

Anonymous
Benjamin Keele
Boycott Elsevier
Brian Hole
cesarberrios
Dorothea Salo
Erin Jonaitis
Ian Davis
jeff
Konrad Förstner
Kristin Briney
Matt Ritter
Neil.Richards
Philip Young
pmlozeau
ranti
Ray Schwartz
rchampieux
Sierra Williams
Stefano Costa
stephaniedawson
zeborah

Editor's Introduction

Samuel A. Moore

Panton Fellow, Open Knowledge Foundation & Dept. Digital Humanities Ph.D Programme, King's College London, London, UK

Panton Fellowships

This book is the result of a year-long Panton Fellowship with the Open Knowledge Foundation and made possible by the Computer and Communications Industry Association. This is the second year that the fellowships have taken place, so far funding five early-career researchers across Europe.

Throughout the year, fellows are expected to advocate for the adoption of open data, centred on promotion of the Panton Principles for Open Data in Science (see below). Projects have ranged from monitoring air quality in local primary schools, to

---

How to cite this book chapter:
Moore, S. A. Editor's Introduction. In: Moore, S. A. (ed.) *Issues in Open Research Data*. Pp. 1–9. London: Ubiquity Press. DOI: http://dx.doi.org/10.5334/ban.a

transparent and reproducible altmetrics, to the Open Science Training Initiative and now this volume on open research data.

In addition to the funding and training fellows receive, the Open Knowledge Foundation is a great network of supportive, like-minded individuals who are committed to the broad mission of increasing openness throughout academia, government and society at large. I strongly encourage anyone eligible to consider applying for a future Panton Fellowship—it has been a very rewarding year.

Panton Principles

> Science is based on building on, reusing and openly criticising the published body of scientific knowledge.
> (Murray-Rust et al. 2010)

In 2009, a group of scientists met at the Panton Arms pub in Cambridge, UK, to try to articulate their idea of what best practice should be for sharing scientific data. The result of this meeting was a first draft of the *Panton Principles for Open Data in Science*, which was subsequently revised and published in 2010.

The Principles are predicated upon the idea that openly sharing one's research data is wholly beneficial to the progression of science. Shared data allows research to be replicated, verified, reused and remixed. But research is competitive and there are perceived disincentives that impact on a researcher's desire or ability to share his or her data. However, the culture of data sharing, and open science more generally, appeals to the collaborative side of the researcher, asking them to consider the discipline in which they work and the progression of science over the narrowly focused desire to maintain ownership of raw data and hence maintain a competitive edge on their colleagues. This is the backdrop against which the Panton Principles were drafted: that sharing data is simply better for science.

The original four authors came from a range of scientific disciplines and backgrounds: Peter Murray-Rust, a chemist from the University of Cambridge; Rufus Pollock, founder of the Open Knowledge Foundation; John Wilbanks, then of Creative Commons and now Sage Bionetworks; and Cameron Neylon, a biophysicist formerly of the Science and Technology Facilities Council and now Advocacy Director at the Public Library of Science. Though each author was an advocate for open science, they disagreed on the best ways to share data to the community. As Cameron recounts in his blog post, Peter above all desired a practical and simple set of rules that publishers could easily adopt to encourage data sharing. There were also disagreements on the application of a share-alike clause to ensure that the products of reused data would remain openly available, though this would be at the expense of interoperability with other forms of data sharing (Neylon 2010).

In the end, the authors decided the best solution would be to recommend that, where possible, data should be released into the public domain. They did this through the creation of four simple principles that should govern the sharing of data. In my opinion these are best read as progressive, with each principle building on the previous one, so that by the end there is a clear sense for how data should be best shared to the community. These principles read as follows:

1. **When publishing data, make an explicit and robust statement of your wishes.**

    This very general point informs the researcher that releasing data into the public domain must be done with the necessary care such that users know the data is in the public domain. For data without such a statement, reuse rights will remain ambiguous and the public domain status of the data could potentially be revoked. As the original principles indicate: 'This statement should be precise, irrevocable, and based on an appropriate and recognized legal statement in the form of a waiver or license.' (Murray-Rust et al. 2010).

2. **Use a recognized waiver or license that is appropriate for data.**
   Building on the previous point, the statement of intent should be in the form of a licence, but one that is appropriate for data. The issue of licencing is complex and will be discussed in great detail through the chapters in this book. However, as a starting point, the authors of the Panton Principles recommend that only licences appropriate for data be used, as opposed to the Creative Commons suite of licences (except CC0), or the GNU General Public Licence or other licences intended for software.

3. **If you want your data to be effectively used and added to by others it should be open as defined by the Open Knowledge/Data Definition—in particular, non-commercial and other restrictive clauses should not be used.**
   Again building on the previous point, appropriate licences should have no needlessly restrictive clauses attached to them. For example, data should not be licensed for non-commercial use only, as this prevents the data being combined with other less restrictively licensed datasets. As the authors explain: 'these [non-commercial]

licenses make it impossible to effectively integrate and re-purpose datasets and prevent commercial activities that could be used to support data preservation' (Murray-Rust et al. 2010).

4. **Explicit dedication of data underlying published science into the public domain via PDDL or CCZero is strongly recommended and ensures compliance with both the Science Commons Protocol for Implementing Open Access Data and the Open Knowledge/Data Definition.**

Finally, the principles arrive at the conclusion that the best licence for releasing data into the public domain is either Creative Commons Zero (CC0) or the Open Data Commons Public Domain Dedication and Licence (PDDL). These licences ensure that data can be reused for commercial purposes, without a legal obligation for attribution (though a social obligation still remains), and ensure maximum interoperability and potential for reuse, in keeping with the 'general ethos of sharing and re-use within the scientific community' (Murray-Rust et al. 2010).

The Panton Principles are founded on the idea that science progresses faster when data can be easily shared and reused throughout the community. Of course, the principles presuppose that the data is already curated to best practice (preserved in a suitable repository, available in a non-proprietary form where possible, etc.). Their intention is simply to recommend the steps you should to take to make your data truly *open*.

The Principles themselves have so far been endorsed by hundreds of scientists worldwide, why not add your signature today at http://pantonprinciples.org/endorse/?

The Book

This book is intended to be an introduction to some of the issues surrounding open research data in a range of academic disciplines. It primarily contains newly written opinion pieces, but also a handful of articles previously published elsewhere (with the authors' permission in each instance). Importantly, the book is meant to start a conversation around open data, rather than provide a definitive account of how and why data should be shared. The book is open access, published under the Creative Commons Attribution License (CC BY), to facilitate further debate and allow the contents to be easily and widely spread. Readers are encouraged to reuse, build upon and remix each chapter; repository managers, data curators and other communities are encouraged to detach and distribute the chapters most relevant to them to their peers and colleagues.

Within the book you will find nine chapters on diverse topics ranging from content mining to drug discovery, to the everyday use of open data in a variety of subjects. A number of issues are inextricably linked to open data, such as data citation, ethics of open data, anonymization, long-term preservation and so forth. All of these issues will be dealt with in various capacities in the ensuing chapters. Finally, the contents have been commissioned so as to strike a balance between the theoretical and the practical—some chapters offer critiques of 'open' approaches or of disciplinary approaches to open data, whilst others contain useful how-to guides for researchers who are new to open data and might not know where to begin.

The book is split broadly in two halves. The first half features pieces on general issues around open data. Peter Murray-Rust and colleagues discuss the legal issues surrounding content mining and offer a manifesto for the 'fundamental rights' of scholars to mine content based on the phrase 'the right to read is the right to mine' (Murray-Rust et al. 2014). Next, in 'The Need to Humanize Open Science', Eric C. Kansa offers a critique of open data, and openness in general, arguing that more attention needs to be focused on the broader institutional structures that govern how research is currently conducted and less on the 'narrow technical and licensing interoperability issues' (Kansa 2014). There are then two previously published pieces by Unni Karunakara and Anthony J. Williams et al. on data sharing within the Médecins Sans Frontières organization and the importance of open data in drug discovery, respectively (Karunakara 2014; Williams et al. 2014).

The latter half of the book features chapters on disciplinary approaches to open data, offering practical advice on data sharing and exploring the subject-specific issues that surround it. Sarah Callaghan's piece offers a comprehensive look at open data in the Earth and climate sciences—barriers and drivers, carrots and sticks, and an insightful case study of one author's personal experience of open data (Callaghan 2014). Tom Pollard and Leo Anthony Celi offer a similarly insightful piece on open data in health care, looking specifically at the delicate balance between patient privacy and open data and how the need to 'do no harm' can be negotiated with the move towards data sharing (Pollard & Celi 2014).

Wouter van den Bos and colleagues then offer their perspective on data sharing in the psychological sciences, making a case for the 'need of a common data sharing policy' that responds to the needs of a discipline that has so far failed to embrace openness in any real sense (van den Bos et al. 2014). Next, Ross Mounce looks at open data in palaeontology, particularly at the complicated state of licensing within the discipline and the need for researchers to use only licences that conform to the Open Knowledge Definition (Mounce 2014). Finally, Velichka Dimitrova describes the Open Economics Principles and the need for all economics data to be 'open by default' to facilitate reproducible research and transparency (Dimitrova 2014).

The book is not meant to be a comprehensive overview of open data and there are of course absences of subjects and viewpoints. However, I do hope the contents are informative, stimulating and, most importantly, help start a conversation around issues in open research data.

Work Cited

Callaghan, S 2014 Open data in the earth and climate sciences. In: Moore, S (ed.) *Introduction to Open Research Data*. London: Ubiquity Press, pp. 89–106.

Dimitrova, V 2014 Open research data in economics. In: Moore, S (ed.) *Introduction to Open Research Data*. London: Ubiquity Press, pp. 141–150.

Murray-Rust, P, Molloy, J C and Cabell, D 2014 Open content mining. In: Moore, S (ed.) *Introduction to Open Research Data*. London: Ubiquity Press, pp. 11–30.

Kansa, E C 2014 The need to humanize open science. In: Moore, S (ed.) *Introduction to Open Research Data*. London: Ubiquity Press, pp. 31–58.

Karunakara, U 2014 Data sharing in a humanitarian organization: the experience of Médecins Sans Frontières. In: Moore, S (ed.) *Introduction to Open Research Data*. London: Ubiquity Press, pp. 59–76.

Mounce, R 2014 Open data and palaeontology. In: Moore, S (ed.) *Introduction to Open Research Data*. London: Ubiquity Press, pp. 151–164.

Neylon, C 2014 The Panton Principles: finding agreement on the public domain for published scientific data. *Science in the*

Open, 22 February 2010. Available at http://cameronneylon.net/blog/the-panton-principles-finding-agreement-on-the-public-domain-for-published-scientific-data/ [Last accessed 6 August 2014].

Murray-Rust, P, Neylon, C, Pollock and R, Wilbanks, J 2010 Panton Principles, principles for open data in science. Available at http://pantonprinciples.org [Last accessed 6 August 2014].

Pollard, T, Celi, L A 2014 Open data in health care. In: Moore, S (ed.) *Introduction to Open Research Data*. London: Ubiquity Press, pp. 129–140.

Van den Bos, W, Mirjam, J and Wulff, D 2014 Open minded psychology. In: Moore, S (ed.) *Introduction to Open Research Data*. London: Ubiquity Press, pp. 107–127.

Williams, A J, Wilbanks, J and Ekins, S 2014 Why open drug discovery needs four simple rules for licensing data and models. In: Moore, S (ed.) *Introduction to Open Research Data*. London: Ubiquity Press, pp. 77–88.

Open Content Mining

Peter Murray-Rust,[*] Jennifer C. Molloy[†] and Diane Cabell[‡]

[*]University of Cambridge and OKFN, Cambridge, UK
[†]University of Oxford and Open Knowledge Foundation, Oxford, UK
[‡]Oxford e-Research Centre, Creative Commons and iCommons Ltd, Oxford, UK

Introduction

As scientists and scholars, we are both creators and users of information. Our work, however, only achieves its full value when it is shared with other researchers so as to forward the progress of science. One's data becomes exponentially more useful when combined with the data of others. Today's technology provides an unprecedented capacity for such data combination.

Researchers can now find and read papers online, rather than having to manually track down print copies. Machines (computers) can index the papers and extract the details (titles, keywords etc.) in order to alert scientists to relevant material. In addition,

How to cite this book chapter:
Murray-Rust, P., Molloy, J. C. and Cabell, D. 2014. Open Content Mining. In: Moore, S. A. (ed.) *Issues in Open Research Data*. Pp. 11–30. London: Ubiquity Press. DOI: http://dx.doi.org/10.5334/ban.b

computers can extract factual data and meaning by "mining" the content.

We illustrate the technology and importance of content-mining with 3 graphical examples which represent the state of the art today (**Figure 1–3**). These are all highly scalable (i.e. can be applied to thousands or even millions of target papers without human intervention. There are unavoidable errors for unusual documents and content and there is a trade-off between precision ("accuracy") and recall ("amount retrieved") but in many cases we and others have achieved 95% precision. The techniques are general for scholarly publications and can be applied to theses, patents and formal reports as well as articles in peer-reviewed journals.

Content mining is the way that modern technology makes use of digital information. Because the scientific community is now globally connected, digitized information is being uploaded from hundreds of thousands of different sources (McDonald 2012). With current data sets measuring in terabytes, it is often no longer possible to simply read a scholarly summary in order to make scientifically significant use of such information (Panzer-Steindel & Bernd 2004; Nsf.gov, 2010; MEDLINE, 2013). A researcher must be able to copy information, recombine it with other data and otherwise "re-use" it to produce truly helpful results. Not only is mining a deductive tool to analyze research data, it is the very mechanism by which search engines operate to allow discovery of content, making connections – and even scientific discoveries – that might otherwise remain invisible to researchers. To prevent mining would force scientists into blind alleys and silos where only limited knowledge is accessible. Science does not progress if it cannot incorporate the most recent findings to move forward.

However, use of this exponentially liberating research process is blocked both by publisher-imposed restraints and by law. These

A:

To a solution of 3-bromobenzophenone (1.00 g, 4 mmol) in MeOH (15 mL) was added sodium borohydride (0.3 mL, 8 mmol) portionwise at rt and the suspension was stirred at rt for 1-24 h. The reaction was diluted slowly with water and extracted with CH2Cl2. The organic layer was washed successively with water, brine, dried over Na2SO4, and concentrated to give the title compound as oil (0.8 g, 79%), which was used in the next reaction without further purification. MS (ESI, pos. ion) m/z: 247.1 (M-OH).

B:

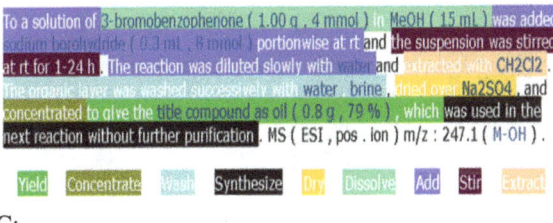

C:

Figure 1: "Text mining". (a) the raw text as published in a scientific journal, thesis or patent. (b) Entity recognition (the compounds in the text are identified) and shallow parsing to extract the sentence structure and heuristic identification of the roles of phrases (c) complete analysis of the chemical reaction by applying heuristics to the result of (b). We have analyzed about half a million chemical reactions in US patents (with Lezan Hawizy and Daniel Lowe).

A:

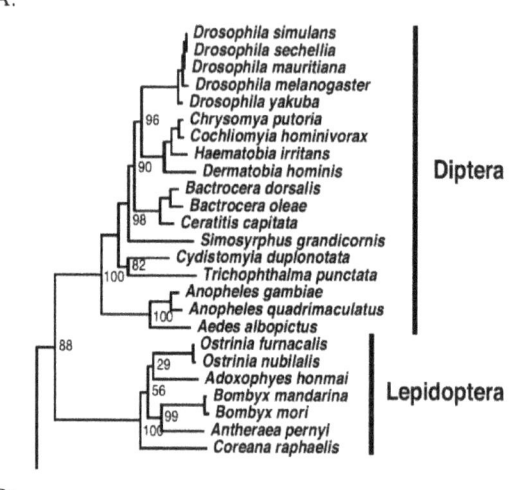

B:
```
<trees label="TreesBlockFromXML" id="Trees" otus="taxl">
 <tree id="tree1" label="tree1" xsi:type="nex:FloatTree">
  <node id="N1" otu="t1" label="Psocoptera "/>
  <node id="N2" otu="t2" label="Sternorrhyncha"/>
  <node id="N3" otu="t3" label="Phthiraptera "/>
  <node id="N4" otu="t4" label="Thysanoptera "/>
  <node id="N5" otu="t5" label="Cicadomorpha "/>
  <node id="N6" otu="t6" label="Heteroptera "/>
  <node id="N7" label="N7"/>
  <node id="N8" label="N8"/>
  ...
  <edge id="line183" label="line183" source="N16" target="N1"/>
  <edge id="polyline176.3" label="polyline176.3" source="N15" target="N2"/>
  <edge id="polyline177.3" label="polyline177.3" source="N14" target="N3"/>
  <edge id="polyline178.1" label="polyline178.1" source="N11" target="N4"/>
  <edge id="polyline180.1" label="polyline180.1" source="N10" target="N5"/>
  <edge id="polyline181.1" label="polyline181.1" source="N7" target="N6"/>
  <edge id="polyline179.1" label="polyline179.1" source="N8" target="N7"/>
  <edge id="polyline175.1" label="polyline175.1" source="N9" target="N8"/>
  ...
 </tree>
</trees>
```

Figure 2: Mining content in "full-text". (a) a typical "phylogenetic tree" [snippet] representing the similarity of species (taxa) – the horizontal scale can be roughly mapped onto an evolutionary timeline; number are confidence estimates and critical for high quality work. These trees are of great value in understanding speciation and biodiversity and may require thousands of hours of computation and are frequently only published as diagrams. (b) Extraction of formal content as domain-standard (NE)XML. This allows trees from different studies to be formally compared and potentially the creation of "supertrees" which can represent the phylogenetic relation of millions of species.

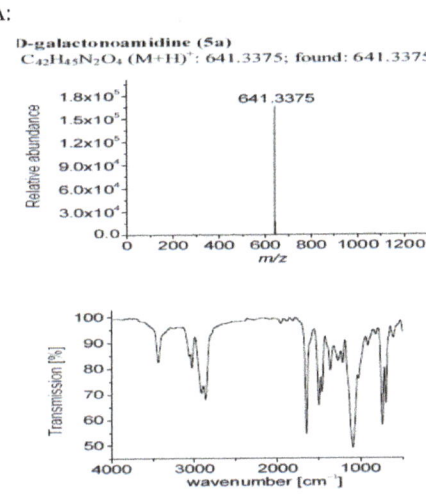

Figure 3: Content-mining from "Supplemental Data" (or "Supporting Information"). This data is often deposited alongside the "full-text" of the journal, sometimes behind the publishers firewall, sometimes openly accessible. It may run to tens or hundreds of pages and for some scientists it is the most important part of the paper. (a) exactly as published [snippet]. Note the inconvenient orientation (designed for printing) and the apparent loss of detail. (b) after content mining techniques and re-orientation – for the "m/z" spectrum (note the fine structure of the main peak, not visible in (a)). It would be technically possible to recover >> 100,000 spectra like this per year from journals.

constraints are based on business models that still rely on print revenue and are supported by copyright laws originally designed for 18th century stationers[1]. While Open Access (OA) practices are improving the ability of researchers to read papers (by removing access barriers), still only around 20% of scholarly papers are offered under OA terms (Murray-Rust 2012). The remainder are locked behind pay walls. As per the terms imposed by the vast majority of journal subscription contracts, subscribers may read pay-walled papers but they may not mine them.

Securing permission to mine on a journal-by-journal basis is extraordinarily time consuming. According to the Wellcome Trust, 87% of the material housed in UK's main medical research database (UK PubMedCentral) is unavailable for legal text and data mining (Hargreaves 2011). A recent study funded by the Joint Information Systems Committee (JISC), an association funded by UK higher education institutions, frames the scale of the problem:

In the free-to-access, UKPMC repository there are 2930 full-text articles, published since 2000, which have the word 'malaria' in the title.

Of these 1,818 (62%) are Open Access and thus suitable for text mining without having to seek permission. However, the remaining 1,112 articles (38%) are not open access, and thus permission from the rights-holder to text-mine this content must be sought.

The 1,112 articles were published in 187 different journals, published by 75 publishers.

---

[1] The Statute of Anne was the first UK law to provide for copyright regulation by government. See Statute of Anne, Wikipedia at http://en.wikipedia.org/wiki/Statute_of_Anne

As publisher details are not held in the UKPMC database, the permission-seeking researcher will need to make contact with every journal. Using a highly conservative estimate of one hour research per journal title (i.e., to find contact address, indicate which articles they wish to text-mine, send letters, follow-up non-responses, and record permissions etc.) this exercise will take 187 hours. Assuming that the researcher was newly qualified, earning around £30,000 pa, this single exercise would incur a cost of £3,399.

In reality however, a researcher would not limit his/her text mining analysis to articles which contained a relevant keyword in the title. Thus, if we expand this case study to find any full-text research article in UKPMC which mentions malaria (and published since 2000) the cohort increases from 2,930 to 15,757.

Of these, some 7,759 articles (49%), published in 1,024 journals, were not Open Access. Consequently, in this example, a researcher would need to contact 1,024 journals at a transaction cost (in terms of time spent) of £18,630; 62.1% of a working year (Hargreaves 2011)

Data and the Law

The intention of copyright law is to support public dissemination of original works so that the public may benefit from access to them. It accomplishes this goal by granting to authors and creators a period of monopoly control over public use of their works so that they might maximize any market benefits. While these principles may work well to protect film producers and musicians, in the current digital environment it is the unfortunate case that they actually delay or block the effective re-use of research results by the scientific community. Research scientists rarely receive any share of the profits on sales of their journal articles, but do benefit greatly by having other scientists read and cite their work. Their interest is therefore best served by maximizing user access and use of their published results.

Databases are protected in a number of ways, most commonly by copyright and database laws. Copyright protects "creative expression" meaning the unique way that an author presents his intellectual output and it prohibits any one from copying, publicly distributing, and adapting the original without permission of the author. Specific statements of facts, shorn of any creative expression as is the case with many types of data, are themselves not ordinarily copyrightable as individual items. However, copyright does come into play for individual data points that exhibit creative expression, such as images (photographs). A collection of data can also be protected by copyright if there is sufficient creativity involved in the presentation or arrangement of the set. In the case of collections, it is only the right to utilize the collection as a whole that is restricted while the individual facts within the collection remain free.

Databases are additionally and independently protected under a *sui generis* regime imposed by the 1996 EU Database Directive (European Parliament 1996). Under the Directive, rights are granted to the one who makes a substantial investment in obtaining, verifying or presenting the contents of the database. Permission of the maker is required to extract or re-utilize all or a substantial portion of the database or to continuously extract or re-utilize insubstantial parts on a continuing basis.

To further complicate matters, copyright and database laws differ from each other and also from one jurisdiction to another. Copyrights may last for more than a hundred years (life of the author plus 70 years). Database rights (which could apply to the self-same database) only run for 15 years however those rights can be extended indefinitely by adding new data to produce a new "work" thus triggering a new term of rights, making it horrendously difficult to determine whether or not protection has expired. The United States, for example, does not impose any *sui generis* rights. Copyright ownership belongs to the creator or his employer, but may be transferred to another (such as a publisher) hence copyright ownership can be difficult to ascertain, particularly where multiple researchers have contributed to the whole. Legal rights in such cases may be jointly held and/or held by one or more employers and/or held by one or more publishers or repositories. The authors of many "orphan" works are unknown or unidentifiable. The more globally-developed the database, the more sets of laws come into play to further complicate the definition of rights.

There are exceptions to such laws when work may be used for specific purposes without permission of the owner. In the UK, these come under the rubric "fair dealing." The UK has a current exception for noncommercial research and private study, however much research is conducted by commercial entities such as pharmaceutical companies.

Even where the law would allow free use of data, publishers imposed restrictions (**Table 1**). The terms of the user's subscription contract – deemed to be a private contract by mutually consenting parties—thus overrides any copyright or database freedoms allowed by law.

Proposed changes in legal policy

Government studies have recognized the harm such restrictions cause to the advancement of science and economic development. They argue that mining is a "non-consumptive" use that does not directly trade on the underlying creative and expressive purpose

Publisher	License Agreement Link	Explicitly prohibits text/data mining?	Quote from standard license agreement
InformaWorld	http://www.informaworld.com/smpp/termsandconditions_partiiintellectualproperty	Yes	"This licence does not include any derivative use of the Site or the Materials, any collection and use of any product listings, descriptions, or prices; any downloading or copying of account information for the benefit of another merchant; or any use of data mining, robots or similar data gathering and extraction tools. In addition, you may not use meta tags or any other "hidden text" utilising our name or the name of any of our group companies without our express written consent.
Taylor Francis	http://www.tandf.co.uk/journals/pdf/terms.pdf	Yes	Incorporates Informaworld terms – see above
Elsevier/CDL	http://orpheus-1.ucsd.edu/acq/license/cdlelsevier2004.pdf	Yes	"Schedule 1.2(a) General Terms and Conditions "RESTRICTIONS ON USAGE OF THE LICENSED PRODUCTS/ INTELLECTUAL PROPERTY RIGHTS" GTC1] "Subscriber shall not use spider or web-crawling or other software programs, routines, robots or other mechanized devices to continuously and automatically search and index any content accessed online under this Agreement."

Blackwell	http://www.blackwellpublishing.com/pdf/Site_License.PDF	No	
OUP	http://www.oxfordjournals.org/help/instsitelicence.pdf	No	
Wiley	http://www.mpdl.mpg.de/nutzbed/wiley-interscience-backfile-co-nutzungs bedingung.pdf	Probably	The systematic downloading of data and the use of excerpts from databases for commercial purposes or for systematic distribution are prohibited.
ACS	http://www.mpdl.mpg.de/nutzbed/MPG_ACS_2002.pdf?la=en	Yes	Licensee (Consortium or Single Institution) acknowledges that ACS may prevent Members and their patrons, as the case may be, from using, implementing or authorizing use of any computerized or automated tool or application to search, index, test or otherwise obtain information from Licensed Materials (including without limitation any "spidering" or web crawler application) that has a detrimental impact on the use of the services under this Agreement.

(Table continued on next page)

(Table continued from previous page)

Publisher	License Agreement Link	Explicitly prohibits text/data mining?	Quote from standard license agreement
AIP	http://www.mpdl.mpg.de/nutzbed/MPG_AIP.pdf	Yes	Systematic or programmatic downloading, printing, transmitting, or copying of the Licensed Materials is prohibited. "Systematic or Programmatic" means downloading, printing, transmitting, or copying activity of which the intent or the effect is to capture, reproduce, or transfer the entire output of a journal volume, a journal issue, or a journal topical section, or sequential or cumulative search results, or collections of abstracts, articles, tables of contents. Other such systematic or programmatic use of the Licensed Materials that interferes with the access of Authorized Users or that may affect the performance of SCITATION, for example, the use of "robots" to index content, or downloading or attempting to download large amounts of material in a short period of time, is prohibited. Redistribution of the Licensed Materials, except as permitted in Section 4, without permission of the Publishers and/or payment of a royalty to the Publishers or to the appropriate Reproduction Rights Organization, is prohibited

BMJ	http://group.bmj.com/group/about/legal/bmj-group-online-licence-single-institution-licence	No	
JSTOR	http://www.jstor.org/page/info/about/policies/terms.jsp	Yes	Prohibited Uses. Institutions and users may not:... f) undertake any activity that may burden JSTORs server(s) such as computer programs that automatically download or export Content, commonly known as web robots, spiders, crawlers, wanderers or accelerators;
Nature	http://www.nature.com/libraries/site_licenses/2010acad_row.pdf	Yes	3. USAGE RESTRICTIONS Except as expressly permitted in Clause 2.1, the Licensee warrants that it will not, nor will it licence or permit others to, directly or indirectly, without the Licensor's prior written consent: (j) make mass, automated or systematic extractions from or hard copy storage of the Licenced Material.

Table 1: Publisher content mining policies.

of the original work or compete with its normal exploitation. Most recently, the 2011 Government-sponsored Hargreaves Report on intellectual property reform, found:

Researchers want to use every technological tool available, and they want to develop new ones. However, the law can block valuable new technologies, like text and data mining, simply because those technologies were not imagined when the law was formed. In teaching, the greatly expanded scope of what is possible is often unnecessarily limited by uncertainty about what is legal. Many university academics – along with teachers elsewhere in the education sector – are uncertain what copyright permits for themselves and their students. Administrators spend substantial sums of public money to entitle academics and research students to access works which have often been produced at public expense by academics and research students in the first place. Even where there are copyright exceptions established by law, administrators are often forced to prevent staff and students exercising them, because of restrictive contracts. Senior figures and institutions in the university sector have told the Review of the urgent need reform copyright to realise opportunities, and to make it clear what researchers and educators are allowed to do. (Hargreaves 2011)

Hargreaves recommended that the Government introduce a UK exception in the interim under the non-commercial research heading to allow use of analytics for non-commercial use, as in the malaria example above, as well as promoting at EU level an exception to support text mining and data analytics for commercial use. It argues that it is "not persuaded that restricting this transformative use of copyright material is necessary or in the UK's overall economic interest." (Hargreaves 2011)

Hargreaves also urged the government to change the law at both the national and EU level to prevent any copyright exceptions from being overridden by contract.

Applying contracts in that way means a rights holder can rewrite the limits the law has set on the extent of the right conferred by copyright. It creates the risk that should Government decide that UK law will permit private copying or text mining, these permissions could be denied by contract. Where an institution has different contracts with a number of providers, many of the contracts overriding exceptions in different areas, it becomes very difficult to give clear guidance to users on what they are permitted. Often the result will be that, for legal certainty, the institution will restrict access to the most restrictive set of terms, significantly reducing the provisions for use established by law. Even if unused, the possibility of contractual override is harmful because it replaces clarity ("I have the right to make a private copy") with uncertainty ("I must check my licence to confirm that I have the right to make a private copy"). The Government should change the law to make it clear no exception to copyright can be overridden by contract" (Hargreaves 2011)

The current U.K. government also believes that the ability for research to power economic development will be greatly enhanced if content mining is encouraged. In responding to Hargreaves, the Government stated its intention to:

- bring forward proposals for a substantial opening up of the UK's copyright exceptions regime, including a wide non-commercial research exception covering text and data mining, and
- aim to secure further flexibilities at EU level that enable greater adaptability to new technologies, and

- make the removal of EU level barriers to innovative and valuable technologies a priority to be pursued through all appropriate mechanisms. (HM Government 2011)

Further, the Government believes that it is not appropriate for "certain activities of public benefit such as medical research obtained through text mining to be in effect subject to veto by the owners of copyrights in the reports of such research, where access to the reports was obtained lawfully." (HM Government 2011)

Because science is a global enterprise, change in copyright law at the national and regional levels will not be sufficient to allow the free flow of information throughout the scientific community. Such changes must be made at many national and regional levels if the goal of a free and open exchange of data is to be achieved.

Changes in publication policies

Because publishers can override legal freedoms by enforcing restrictive terms of use in subscription agreements, we urge researchers to not only support these Government initiatives, but to go further by taking personal and institutional responsibility for establishing open mining practices in their work and publishing environments. In particular, we urge the adoption of the following Open Mining Manifesto (Murray-Rust 2012).

Open Mining Manifesto

1. Define 'open content mining' in a broad and useful manner

'Open Content Mining' means the unrestricted right of subscribers to extract, process and republish content manually or by machine in whatever form (text, diagrams, images, data, audio, video, etc.) without prior specific permissions and subject only to community norms of responsible behaviour in the electronic age.

[1] Text
[2] Numbers
[3] Tables: numerical representations of a fact
[4] Diagrams (line drawings, graphs, spectra, networks, etc.): Graphical representations of relationships between variables, are images and therefore may not be, when considered as a collective entity, data. However, the individual data points underlying a graph, similar to tables, should be.
[5] Images and video (mainly photographic)- where it is the means of expressing a fact.
[6] Audio: same as images – where it expresses the factual representation of the research.
[7] XML: Extensible Markup Language (XML) defines rules for encoding documents in a format that is both human-readable and machine-readable."
[8] Core bibliographic data: described as "data which is necessary to identify and / or discover a publication" and defined under the Open Bibliography Principles [15].
[9] Resource Description Framework (RDF): information about content, such as authors, licensing information and the unique identifier for the article.

2. Urge publishers and institutional repositories to adhere to the following principles:

Principle 1: Right of Legitimate Accessors to Mine

We assert that there is no legal, ethical or moral reason to refuse to allow legitimate accessors of research content (OA or otherwise) to use machines to analyse the published output of the research community. Researchers expect to access and process the full content of the research literature with their computer programs and should be able to use their machines as they use their eyes. **The right to read is the right to mine**

Principle 2: Lightweight Processing Terms and Conditions

Mining by legitimate subscribers should not be prohibited by contractual or other legal barriers. Publishers should add clarifying language in subscription agreements that content is available for information mining by download or by remote access. Where access is through researcher-provided tools, no further cost should be required. **Users and providers should encourage machine processing**

Principle 3: Use

Researchers can and will publish facts and excerpts which they discover by reading and processing documents. They expect to disseminate and aggregate statistical results as facts and context text as fair use excerpts, openly and with no restrictions other than attribution. Publisher efforts to claim rights in the results of mining further retard the advancement of science by making those results less available to the research community; such claims should be prohibited. **Facts don't belong to anyone.**

3. Strategies

Assert the above rights by:

- Educating researchers and librarians about the potential of content mining and the current impediments to

doing so, including alerting librarians to the need not to cede any of the above rights when signing contracts with publishers
- Compiling a list of publishers and indicating what rights they currently permit, in order to highlight the gap between the rights here being asserted and what is currently possible
- Urging governments and funders to promote and aid the enjoyment of the above rights.

Editor's note

This article originally was originally presented at the Conference for the Fellows of OpenForum Academy, 24th September 2012 in Brussels, and is reproduced in accordance with the CC BY licence and with kind permission of the authors. Whilst there have been no alterations to the content, the reference style has been amended for consistency with the other chapters in the book.

References

European Parliament (1996) *Directive 96/9/EC of the European Parliament and of the Council of 11 March 1996 on the legal protection of databases.* Available at at http://eur-lex.europa.eu/LexUriServ/LexUriServ.do?uri=CELEX:31996L0009:EN:HTML [accessed 22 Sep. 2014].

Hargreaves (2011) *Digital Opportunity, A Review of Intellectual Property and Growth* available at http://www.ipo.gov.uk/ipreview-finalreport.pdf [accessed 22 Sep. 2014].

HM Government (2011) *The Government Response to the Hargreaves Review of Intellectual Property and Growth.* Available at http://www.ipo.gov.uk/ipresponse-full.pdf.

Panzer-Steindel, Bernd (2004) *Sizing and Costing of the CERN T0 center*, CERN-LCG-PEB-2004-21 available at: http://lcg-computing-fabric.web.cern.ch/lcg-computing-fabric/presentations/Sizing%20and%20costing%20of%20the%20CERN%20T0%20center.doc [accessed 22 Sep 2014].

McDonald, (2012) The Value and Benefits of Text Mining, Section 3.3.8, JISC Report Doc #811, available at http://www.jisc.ac.uk/publications/reports/2012/value-and-benefits-of-text-mining.aspx.

MEDLINE (2014). *MEDLINE® Citation Counts by Year of Publication (as of mid – November 2013)*. [online] Available at: http://www.nlm.nih.gov/bsd/medline_cit_counts_yr_pub.html [Accessed 22 Sep. 2014].

Murray-Rust P (2012) *The Right to Read is the Right to Mine* http://blog.okfn.org/2012/06/01/the-right-to-read-is-the-right-to-mine/ [accessed 22 Sep 2014].

Nsf.gov, (2010). *nsf.gov – Science and Engineering Indicators 2010 – Chapter 5. Academic Research and Development – Highlights – US National Science Foundation (NSF)*. [online] Available at: http://www.nsf.gov/statistics/seind10/c5/c5h.htm [Accessed 22 Sep. 2014].

Open Knowledge Foundation (2011) *Principles on Open Bibliographic Data*. Available at http://openbiblio.net/principles/ [Accessed 22 Sep. 2014].

The Need to Humanize Open Science

Eric C. Kansa
OpenContext.org

Introduction

The "open science" movement has reached a turning point. After years of advocacy, governments and major granting foundations have embraced many elements of its reform agenda. However, despite recent successes in open science entering the mainstream, the outlook for enacting meaningful improvements in the practice of science (and scholarship more generally) remains far from certain.

The open science movement needs to widen the scope of its reform agenda. Traditional publishing practices and modes of conduct have their roots in institutions and ideologies that see

How to cite this book chapter:
Kansa, E. C. 2014. The Need to Humanize Open Science. In: Moore, S. A. (ed.) *Issues in Open Research Data*. Pp. 31–58. London: Ubiquity Press. DOI: http://dx.doi.org/10.5334/ban.c

little critique among proponents of open access and open data. A focus solely on the symptoms of dysfunction in research, rather than the underlying causes, will fail to deliver meaningful positive change. Worse, we run the risk of seeing the cause of "openness" subverted to further entrench damaging institutional structures and ideologies. This chapter looks at the need to consider openness beyond narrow technical and licensing interoperability issues and explore the institutional structures that organize and govern research.

Background

I am writing this contribution from the perspective of someone actively working to reform scholarly communications. I lead development of Open Context, an open data publishing venue for archaeology and related fields.[1] Like many such efforts, most of Open Context's funding comes from grants. Much of my time and energy necessarily goes toward raising the money needed to cover staffing and development costs associated with giving other people's data away for free. My struggles in promoting and sustaining open data inform the following discussion about the institutional context of the open science movement.

My own academic training (doctorate in archaeology) straddles multiple disciplinary domains. Few universities in the United States have departments of "archaeology." Instead, archaeology is taught in departments of anthropology (as in my case), classics, East Asian studies, Near Eastern studies, and other programs of humanities "area studies." Within archaeology itself, many researchers see themselves first and foremost as scientists

---

[1] http://opencontext.org

attempting to document and explain economic, ecological, and evolutionary changes in human prehistory, while others orient themselves more toward the humanities, exploring arts, ideologies, identity (gender, ethnicity, class, etc.), spirituality, and other aspects of the lived experience of ancient peoples. Most archaeological field research, whether it emphasizes "scientific" or "humanistic" research questions, involves inputs from a host of specializations from many different fields. Archaeologists routinely need to synthesize results from a vast range of disciplines, such as geological sciences, material science and chemistry, zoology, botany, human physiology, economics, sociology, anthropology, epigraphy, and art history.

Humanities and Open Science

The wide interdisciplinary perspective of my background in archaeology makes me uncomfortable with some of the rhetoric of open science. From the perspective of an archaeologist, the "science" part of open science is not only vague, but seems to privilege only one aspect of our research world. The divide between what is and what is not considered to be science harkens back to historical contingency and institutional and political structures that allocate prestige and finances. In the US, science involves research activities funded by the National Institutes for Health (NIH) and the National Science Foundation (NSF). Other research interests lie at the margins and receive significantly less public support. Vested interests give these institutional structures a great deal of inertia and make them hard to change.

Digital technologies, data, data visualization, statistical analyses, and sophisticated semantic modeling now lie at the heart of many areas of humanistic study, often lumped together as the "digital humanities." Digital humanities research, like many areas of scientific research, also increasingly emphasizes access, reduction of intellectual property barriers, reproducibility, transparent algorithms, wide collaboration, and other hallmarks of open science. In other words, humanists and digital humanists often care as deeply about issues of intellectual rigor, application of appropriate theoretical models, and the quality of evidence as their lab-coat-wearing colleagues. Indeed, two (William Noel and myself) of the ten "Champions of Change" recognized by the White House in 2013 for contributions in open science were primary funded by the National Endowment for the Humanities Office of Digital Humanities (NEH-ODH).[2] This is a remarkable achievement for the digital humanities community, considering that the entire NEH only sees a budget of US$140 million per year—orders of magnitude less than both the NIH (US$20 billion per year) and the NSF (US$2 billion per year).

It is very difficult and arguably damaging to draw sharp boundaries in research so as to define science in opposition to other areas of inquiry. Archaeology is just one area where such boundaries routinely blur. The rise of the Digital Humanities does not necessarily mean an encroachment of scientific perspectives and methods into rather more interpretive and mathematics-shy areas of cultural study. Some of the discussion surrounding "Culturomics," a term coined by Erez Aiden and Jean-Baptiste Michel to give their analyses of Google Books data (Michel et al. 2011) the same sort of scientific cachet as "genomics" or "proteomics," implies a sort of triumph of statistically powered empiricism over

---

[2] See: http://www.neh.gov/divisions/odh/featured-project/neh-grantees-honored-white-house-open-science-champions-change

the pejoratively fuzzy, subjective, and obtuse humanities (see a fascinating discussion of this in Albro 2012).

The fact that we now have large datasets documenting cultural phenomena will not automatically transform humanistic research into just another application area of Data Science. A key research focus of the humanities (and many social sciences) centers on critique and analysis of otherwise tacit assumptions and *a priori* understandings. Like any area of intellectual inquiry, critique can be done badly, and there are plenty of examples of humanistic critique that read like self-parody. Nevertheless, humanities and social sciences perspectives can offer powerful insights into science's institutional and ideological blind spots, including the blind spots of open science.

With these issues in mind, I will continue to use the phrase "open science" in this discussion. However, the "science" I discuss refers to a wider universe of systematic study than often considered in contemporary university or policy-making bureaucracies. My use relates more to the Latin root of the term, *scientia*, referring to knowledge, or the German word *Wissenschaft*, signifying scholarship involving systematic research or teaching.[3] I am adhering to the language of open science to help make sure the humanities, including the digital humanities, are part of the conversation on how we work to reform research more generally.

Open Science and "Conservatism"

Many academic researchers, at least in archaeology, the field I know best, are still largely oriented toward publication expectations

---

[3] Thanks to @openscience for helping me explore these issues. I am very gratified by the commitment of @openscience toward all areas of research, including the humanities and social sciences.

rooted in mid-20th or even 19th century practice. But that orientation does not reflect our current information context. The World Wide Web has radically transformed virtually every sphere of life, including our social lives, commerce, government, and, of course, news, entertainment, and other media.

The Web itself grew out of academia, as a means for researchers at CERN and other university laboratories to efficiently share documents. Ironically, academia has been remarkably reluctant to fully embrace the Web as a medium for dissemination. The humanities and social sciences, including archaeology, are notable in how little social and intellectual capital we invest in web-based forms of communication.

The reluctance of many academics to experiment with new forms of scholarly communication stands as one of the central challenges in my own work with promoting data sharing in archaeology. One would naively think that data sharing should be an uncontroversial "no-brainer" in archaeology. After all, archaeological research methods, particularly excavation, are often destructive. Primary field data documenting excavations represent the only way excavated (i.e. destroyed) areas can ever be understood. One would think this would make the dissemination and archiving of primary field data a high priority, particularly for a discipline that emphasizes preservation ethics and cultural heritage stewardship (Kansa 2012).

Despite these imperatives, archaeologists often resist or avoid investing effort in data stewardship. It may be tempting to cite academic conservatism as a rationale for this reluctance, but this has little explanatory power. Archaeologists are, if anything, very selective in their "conservatism." Many are highly engaged with new technologies. Photogrammetry (sophisticated digital image processing), X-ray defraction (instruments to study chemical compositions), geographic information systems, remote sensing (satellite and other reconnaissance data), various geophysical methods (ground penetrating radar, magnetometry), three-dimensional modeling, and even drones see rapid adoption in the discipline. Archaeologists also have professional incentives to distinguish themselves among their peers and do so through publishing innovative approaches in archaeological methods, theories, or interpretations. However, while archaeologists strive to innovate in many areas of their professional lives, publication practices remain highly resistant to change. To explore why, we need to look at the larger institutional and professional context in which academic archaeologists work. This context is broadly similar to many other areas of research and can help illuminate issues faced in promoting open science.

Open Context, Open Data, and Publication

Publication lies at the heart of most fields of academic inquiry. It plays an integral role in our success in finding grants and employment, and it helps structure our identities as researchers. The economics, expectations, and constraints of publishing practices help shape what we know and communicate in all fields of research. In the case of archaeology, the communication and preservation of primary field data and documentation fits poorly into normative publishing practices. This leads directly to the hoarding, neglect, and loss of archaeological data.

Many of our colleagues prioritize publication goals over virtually every other professional goal. We have to understand and negotiate this reality in our efforts to promote data sharing in archaeology. To this end, Open Context, the data sharing system

I direct, has adopted a model of "data sharing as publication." Open Context publishes a wide variety of archaeological data, ranging from archaeological survey datasets to excavation documentation, artifact analyses, chemical analyses of artifacts, and detailed descriptions of bones and other biological remains found in archaeological contexts. The datasets comprise rich media collections, including tens of thousands of drawings, plans, and photos of artifacts, archaeological deposits, and ancient architectural features. The range, scale, and diversity of these data require dedicated expertise in data modeling and a sustained commitment to continual development and iterative problem solving. Most content in Open Context carries a Creative Commons Attribution License and can be retrieved in a variety of machine-readable formats (XML, CSV, JSON, RDF).

We use "data sharing as publishing" to help encapsulate and communicate the investment and skills needed for sharing reusable data. A publishing metaphor can help put that effort into a context that is readily recognized by the research community (i.e. data publishing implies efforts and outcomes similar to conventional publishing). We hope offering a more formalized approach to data sharing will promote professional recognition (as noted by Harley et al 2010), which would motivate better data creation practices at the outset. Ideally, "data sharing as publishing" can help create the reward structures that make data reuse less costly and more scientifically rewarding (Kansa & Kansa 2013). Open Context uses the EZID system to mint persistent identifiers (digital object identifiers (DOIs) and archival resource keys (ARKs), and archives data with the University of California's California Digital Library, a unit that runs a major digital repository called Merritt. Archiving and persistent identifiers provide a stable foundation for the citation of data, an important issue to consider in situating data sharing within the Academy's conventions and traditions (see also Costello 2009).

At the same time, we recognize some of the limits of using "publication" as a metaphor for data sharing. In our experience publishing data, some problems in data recording and documentation only became evident after researchers actually tried to reuse and analyze each others' datasets (Atici et al. 2013; Kansa, Whitcher Kansa & Arbuckle 2014). In other words, problems in a dataset may go undetected even after cycles of editorial review and revision, only to be discovered long after publication. Even using the term publication with data can carry the unfortunate baggage of implying finality or fixity.

Open Context's datasets are not fixed as static products, despite our use of the term "publication". For instance, we need to revise datasets periodically to fix errors or to annotate with new controlled vocabularies and ontologies through linked open data methods. In many respects, then, Open Context treats datasets as software source code. Like source code, the data are expressed as structured text and new versions are "pushed" to the community of users. The use of version control systems (such as GitHub, in the case of Open Context) can improve the management, professionalism, and documentation associated with ongoing and collaborative revision of datasets (for a thoughtful discussion see Kratz and Strasser 2014).[4]

Open Context now publishes key datasets in a number of specializations covering topics as diverse as the development of early agricultural economies to the comprehensive settlement history

---

[4] A recent paper details a case study for how Open Context's editorial, annotation, publishing, and version control practices assisted in the analysis and interpretation of multiple datasets submitted by 34 archaeologists studying early agriculture in Turkey (Kansa et al. 2014).

of large portions of North America.[5] Despite these developments, data sharing and data publication are still not expected aspects of archaeological scholarship. Though data management sees growing recognition in the archaeological community, few archaeologists feel free to commit to the effort required to improve data quality and documentation. This hesitation stems from relentless professional pressures that make any deviation from established norms almost unthinkable.

A Context of Neoliberalism

Why is it so difficult for many researchers to deviate from established modes of publication? This question lies at the heart of many discussions about open science and scholarly communications. And while most open science advocates acknowledge the challenge of overcoming professional incentives that inhibit reform, there has not been enough discussion of the institutional basis of those (dysfunctional) professional incentives.

In much of the wealthy, industrialized world, the past three decades have witnessed an accelerating consolidation of "neoliberalism," a loosely-associated set of ideologies, economic policies, and institutional practices. Using a vaguely defined term like neoliberalism can be problematic especially when applied too broadly (Kingfisher & Maskovsky 2008). However, this discussion focuses on policy and governance issues in academic institutions, and in

---

[5] The Digital Index of North American Archaeology (DINAA project, led by David G. Anderson and Joshua Wells) publishes "site files" compiled by state government officials that inventory and document archaeological and historical sites identified by archaeologists. Identification of most of these archaeological sites resulted from contracted studies ("cultural resource management") to comply with federal historical protection laws. See: http://ux.opencontext.org/blog/archaeology-site-data/

this context, neoliberalism offers a useful shorthand for discussing a variety of loosely related ideologies and practices (Lorenz 2012; Feller 2008). Very broadly, neoliberalism refers to policies of economic liberalization (deregulation), imposition of "market-based" dynamics (as opposed to central planning or public support and financing), and corporate management methodologies, especially workplace monitoring and performance incentives.

What does neoliberalism have to do with academic publishing? As it turns out, virtually everything about scholarly publishing in one way or another relates back to neoliberal policy making. Over the past few decades, consolidation in academia's commercial publishers has helped fuel dramatic price increases, averaging 7.6% per year for the past two decades and amounting to 302% cost increases between 1985 and 2005 (McGuigan et al. 2008). Ideologies and policies favoring market deregulation permit such commercial consolidation. At the same time, escalating subscription costs give commercial publishers consistently high profit margins—35% in the case of Elsevier (Mobbiot 2011). These price increases further exacerbate other outcomes of neoliberal policies. While publication costs skyrocket, academic libraries witness declining budgets as higher education institutions struggle in a climate of fiscal austerity.

The escalating cost of higher education, or, rather, the increasing co-option of research and educational funding streams toward corporate interests, inevitably means that academic institutions pass the costs of neoliberal policies on to core constituents, namely faculty and students. Researchers see reduced salaries, smaller research budgets, and cut-throat competition for fewer faculty positions. Academic labor has become increasingly contingent, as part-time and short-term adjunct faculty contracts have become the norm. The pay and working conditions of this contingent class of scholars requires many of them to supplement their income with public welfare assistance to pay for basic necessities such as food and shelter (Patton 2012). At the same time, students see explosive growth in tuition, and in the US, this has fed a mind-boggling US$1.2 trillion level of student debt (Denhard 2013).

Neoliberal pressures on archaeological publication extend beyond cost increases and reduced public financing. Neoliberal ideologies also emphasize "instrumentalism" in research and education (Hamilakis 2004). Policy makers increasingly expect direct and immediate financial returns for investment in education and research. Research, instructional, and other scholarly activities increasingly need to "pay for themselves." This driver makes it extraordinarily difficult to finance open data, especially in a long-term and sustainable manner (see below).

Instrumentalism creates pressure to align scholarship toward easily commercialized ends. Students, under pressure to justify high levels of debt, feel compelled to focus on subjects thought to give high financial returns. Archaeologists often need to justify their course offerings in how they give students "transferrable skills" that can be applied in more practical domains. University administrators also increasingly use instrumentalist rhetoric to argue against further erosion of public financial support. Because most public financing of research goes toward medical, engineering, or scientific domains critical for economic competitiveness (another neoliberal trope), university administrations prioritize these easily monetized domains in new hires, facilities, and other supports. For the humanities and social sciences, including archaeology, this has exacerbated the bite of publication cost escalations. The worst publication cost escalations have focused on science, technology, engineering, and medicine (STEM) journals, yet those are the journals prioritized in library budgets because of their strategic importance to universities. This leaves even less money for books and journals in the humanities and social sciences (Steele 2008; Davidson 2013).

Academic publication is not just about communicating with one's peers. It involves a selection of venues, choice of language and style, and other signals that communicate one's claims to a certain professional identity. Many of us who have taught undergraduate and graduate students have personally observed and mentored student learning in how to communicate like one of us. It is a central aspect of the reproduction of academic culture. The mastery of publication practices can make or break a career, because publication is so heavily invested with prestige and social capital. Journals can have very competitive review processes and rejection rates. A citation or a positive review from an elite scholar has implications for employment. The adage "publish or perish" captures these high stakes.

The Public Library of Science (PLOS) achieved remarkable early success in drawing social capital to its titles. Sadly, the success of PLOS has been slow to replicate in many other disciplines. It is very difficult to promote new and unfamiliar forms of scholarly contribution with uncertain rewards when many in the research community feel increasing pressure to perform in clearly recognized ways. Diane Harley and colleagues (2010) led the largest and most comprehensive investigation of scholarly communications practices to date. Part of their study focused on archaeology. Unsurprisingly, they noted how professional incentives and rewards deter many faculty from participating in digital publishing. Faculty often feel wary of committing effort toward digital projects when mainstream publication offers much more clear and certain rewards.

Counting Publication with Performance Metrics

It is ironic that even though many researchers are reluctant to share data, most common publication incentive structures *treat their papers as data*. In many institutions, hires, promotion, and tenure all center on numeric assessments of a given researcher's publication record.

The growing importance of performance metrics further fuels the competitive fire of academic publishing. The rise of performance metrics represents an important change in academic administration and is often seen as another manifestation of neoliberalism (see overview by Feller 2008). Performance metrics have assumed greater importance in administration and governance because of their apparent objectivity in assessment. Administrative bureaucracies tend to promote metrics because they promote "accountability" by giving clear and quantified outcomes of work these bureaucracies manage and finance. The apparent objectivity of quantification further legitimizes allocation of resources based on metrics. Metrics, unlike more qualitative assessments, seem (at first glance) less susceptible to biasing by age, class, gender, race, or other social and political factors that may color judgments about performance.

Thus, performance metrics are integral aspects of rational meritocracies including, especially, the Academy. One does not need to look hard for examples of how performance metrics help shape academic practice. The UK and Australia have enacted two of the most prominent and ambitious programs of academic performance monitoring with, respectively, the Research Excellence Framework (REF) and the Excellence in Research for Australia (ERA). While the US has a far more decentralized institutional context for universities and has no equivalent to the REF or ERA systems, various performance metrics also feature prominently in allocating resources, at both the institutional and researcher levels (Feller 2008).

In describing metrics, I use the phrase "apparent objectivity" quite deliberately. We live in a vastly complex social world. This complex reality offers many phenomena that we can potentially choose to count and measure. However, even in an era where data collection is cheaper and easier than ever, we select only tiny slices of our overall social reality to quantify. Our models of how people and organizations perform, practical and legal issues, as well as institutional and ideological factors, all shape which social phenomena we choose to measure. These factors come together to make the quantification of a complex social process like research less objective than it can initially seem.

When metrics become significant factors in attracting or allocating financial resources, the choices involved in selecting metrics necessarily become political choices. Metrics measure what certain institutions value, and those measurements can become increasingly valued by institutions. In these circumstances, feedback loops can entrench certain metrics into becoming significant institutional or organizational goals unto themselves. As already discussed, neoliberal policies ratchet up competition for jobs and funding. In this relentlessly competitive context, various institutional and individual performance metrics can become potent motivators toward certain kinds of behavior.

The role of metrics in shaping publication practices has received a great deal of attention. The often criticized Impact Factor started out as a way for librarians to make more informed choices about journal subscriptions (at least according to Curry 2012). In that context, the Impact Factor was relatively benign (see Garfield 2005). However, according to many scientists and other observers (see below), the Impact Factor evolved into a proxy for assessing the quality of individual research contributions. In other words, it became a tool for Taylorism. Taylorism refers to Fredrick Taylor, the originator of "scientific management," a highly influential approach to workplace administration that emphasizes achievement of discreet, quantified goals to promote productivity. It carries negative connotations of coercive monitoring and dysfunctional misalignments between meaningful but abstract goals and the actual behaviors being measured. Above all, Taylorism implies reduced workplace autonomy, diminished creativity, and the dreary mass-production of standardized, readily quantified products. These are precisely the criticisms levied against university bureaucracies that draw on the Impact Factor for hiring and promotion decisions.

Given the potent role played by publication metrics and the difficulty inherent in distilling complex social realities into simple measurements, metrics are hotly debated. In 2013, the San Francisco Declaration on Research Assessment (DORA) was signed by several journal publishers and editors, hundreds of organizations, and, more notably, more than 10,400 members of the scientific community.[6] DORA represents one of the most visible acts of protest against the use of Impact Factors in measuring the quality of an individual's research.

Further demonstrating dissatisfaction with conventional citation metrics, ImpactStory.org recently launched (with major funding from the Alfred P. Sloan Foundation) an effort to provide alternative measurements of research outputs better aligned with Web-based modes of communication. Conventional citation metrics only count papers published in traditional

---

[6] See: http://am.ascb.org/dora/

peer-reviewed journals. Any form of scholarly contribution that falls outside of these venues, such as software, computational models, data, or blog posts literally do not "count." Researchers and institutions that value such alternative forms of scholarship often want their Web-native forms of contributions to count, provoking widespread enthusiasm among reform activists for the type of "altmetrics" (alternative metrics) being developed by ImpactStory.org.

Should We Count on Better Metrics to Make Science Open?

I do not have the expertise to more fully explore issues in bibliometrics (the studies involving the quantification of research publications), nor to discuss the relative merits of different forms of citation analysis and impact rankings. I also do not want to dismiss the field of bibliometrics (or even the Impact Factor itself) as nothing more than a dystopian tool of neoliberalism and Taylorist surveillance. Bibliometrics can be useful and powerful tools in library and information science to promote information discovery, identify linkages between concepts, and other important (from a research perspective) ends. Thus, there is nothing inherently wrong with exploring and refining new types of citation metrics and altmetrics. In fact, this is an important area of research deserving attention and support.

The problem with metrics lies not in quantifying research outputs *per se*, but rather how institutions use metrics to shape behaviors. The clearest problem I see in relying on metrics as a tool for reform centers on the inertia behind the institutionalization of a particular metric. Data sharing advocates often talk about how data should be rewarded just like other forms of publication. Data should "count" with measurable impacts. If we convince universities to monitor data citation metrics, they can incentivize more data sharing. We can also collect a host of altmetrics to incentivize other forms of Web-based collaboration and open source projects.

Unfortunately, it takes a great deal of time to convince university bureaucracies and granting foundations to adopt a new system of metrics. Entrenched constituencies inevitably have vested interests in already established means of assessment. Introducing new metrics that may disrupt an established *status quo* will be a slow and sometimes painful process. By the time a given metric becomes incorporated into administrative structures, the behaviors it tries to measure will not necessarily be innovative anymore. Worse, even the most forward-looking current altmetrics cannot anticipate (or, thus, accommodate) future innovative approaches to scholarly communication. Thus, unanticipated innovations in the future *still* will not count.

Using metrics implies that the objects being measured are commensurate and this may undermine the value of scholarship. For example, a certain dataset may uniquely and irreplaceably document a key epigraphic corpus of a long-dead civilization whose written language is only understood by a dozen scholars worldwide. This dataset may count for next to nothing using conventional impact metrics or even altmetrics. Yet, it would be measured in the same way as a paper describing a new readily commercialized nano-material or a dataset documenting social networks among corporate board members. These different forms of scholarly contribution each have great value in their own right, but their significance is highly context-dependent. It is very difficult to compare their relative worth, and indeed such comparison may cheapen their value in unforeseen ways.

If we see all forms of scholarship as assessable through a common set of metrics, we risk ignoring key contextual associations that differentiate meaningful "knowledge" from mere "data." Ignoring context can mean any given metric will be as arbitrary and meaningless in a given situation as a measure of file size or a paper's alphabetical ranking by title. In other words, there is a danger institutions may use metrics to treat research outputs and data as somehow "fungible" (functionally interchangeable) and in the process devalue or diminish scholarly context.

Both conventional metrics and altmetrics attempt to measure "impact." The website RetractionWatch.com, a venue for tracking increasing levels of publication retraction, notes how an incentive structure favoring quantity and "splashy" findings encourages shoddy research and sometimes outright fraud (see also Fang and Casadevall 2014 noting a strong positive correlation between journal Impact Factor and retraction rates).[7] It is possible there may be even more insidious issues in emphasizing performance metrics and altmetrics that measure impact. Do impact metrics exacerbate "hype-cycles" and band-wagon effects of chasing short-term popularity at the expense of long-term (possibly more meaningful) research programs (Field 2013)? Many forms of impact can be diffuse and difficult to observe, especially when they relate to policy making. I have helped set data sharing agendas for professional societies and granting foundations and none of those activities would count in any conceivable metric or altmetric. I raise these issues because I suspect that we take far too much for granted when we discuss and attempt to measure impact.

---

[7] See this fascinating discussion thread: http://retractionwatch.com/2014/04/07/pain-study-retracted-for-bogus-data-is-second-withdrawal-for-university-of-calgary-group/#comment-90374

Impact is only one of a wide array of possible ways to quantify research. Tim McCormick started a provocative thread on Twitter (McCormick 2014) under the #allmetrics hash-tag, making a clear reference and unique twist to the #altmetrics hash-tag. In the thread, McCormick asks if there are other valences and dimensions to scholarship that can and should be counted than those that measure exposure and attention, as is the case with conventional citation metrics and altmetrics. His comments point to the political processes and ideological assumptions inherent in how certain metrics gain institutional power. One can imagine a whole host of metrics aimed at measuring labor conditions and hiring equity of laboratories publishing biomedical research, or metrics counting investments in mentorship associated with faculty and student field work and data collection. These examples seem almost comical because it is difficult to imagine contemporary universities caring about such issues sufficiently to actually develop policies based on such radically different metrics.

The problems we encounter in encouraging more open, transparent, and collaborative forms of research stem not merely from the reign of certain bad legacy metrics, but from institutional structures that promote profound power inequalities. Those power relationships make metrics far too influential in shaping research agendas, outcomes, and careers. It is the obsession with performance metrics itself, not the choice of metrics, which stifles academic freedom. Researchers need the space and autonomy to experiment, creatively play, take risks, and occasionally fail. The constant pressure to maximize measurable performance inhibits precisely those aspects of science and research we should most value.

Institutional hierarchies are partially defined by who measures and monitors whom, and according to what metric. In other words, establishing and enforcing metrics can be political tools to discipline members of a community. Neoliberal policy seems to care little about the human costs and creativity loss associated with maximizing research productivity as narrowly defined by a given metric. So while altmetrics that incentivize behaviors like data sharing can conceivably gain some traction (after much struggle) in current institutional settings, other more radical forms of "allmetrics" that measure such issues as labor conditions in research are probably nonstarters.

This last point raises an important issue. An open science reform agenda needs to extend beyond a focus on copyright licenses, access to research data, and collaboration on GitHub. Institutionalizing meaningful open science reforms probably also requires reform and reconfiguration of the institutions in which researchers work. Homogeneous career options, institutional structures, and performance metrics will continue to promote homogenous researchers and research outputs. If we want to encourage more innovation and diversity in the conduct of research, we should encourage and reward more diversity in career paths and institutional structures. Innovation in open science will require investing in new institutional forms that better recognize and reward collaboration and communication of the research *process*, not just the finished product.

Though the above discussion highlights my skepticism of using better metrics to "count" our way to open science, recognizing such issues helps us seek alternative approaches. Efforts like ImpactStory.org are important and relevant because they start a much needed conversation about how to encourage higher quality, more collaborative, and more ethical conduct. Yet we should remember that altmetrics need to be the start of the conversation, not the end. The need for reform goes far deeper than selecting the right impact measurements.

Open Science, Public Goods, and Communities

The largest and most entrenched policy barrier to promoting open science centers on the current neoliberal climate of relentless competition. Open science seeks to improve research practice by making the process of research as evident and open for collaboration, scrutiny, reuse, and improvement as the final products of research. Exposing the research process to a wider community requires a high level of collegiality and trust. Cynically, I suspect that collegiality and trust are precisely the personality traits and inclinations that are most at odds with career success in many academic departments.

Any research career now involves tremendous risks ranging from dismal serfdom as an adjunct to complete ejection from academia. Most researchers (save for an exceptionally brave or foolhardy few) are loath to expose themselves to even more risk by adopting novel open science modes of practice. If research remains a hyper-competitive, zero-sum game, no amount of data citation or altmetrics will lead to trust or collaboration. Worse, we could face a situation where counts of datasets and GitHub updates succeed in "open washing," a system whose fundamentals breed anxiety, suspicion, and escalating pressures to cut corners.[8]

The risks of open washing are real. In our efforts to promote open data and open science more generally, we often use neoliberal policy arguments. We emphasize how open data and open science will reduce overall costs and introduce new commercial opportunities for entrepreneurs and their investors. After all,

---

[8] "Open washing" borrows from the phrase "green washing." Green washing describes superficial measures to give the appearance of environmental sound and sustainable practices. Open washing similarly describes superficial and insubstantial measures that signal openness.

canonical definitions of open data require data to be freely available for commercial use without restriction. Awkwardly, someone still needs to finance the creation and maintenance of the open data that can have such wonderful commercial utility. Where will that money come from? On this issue, open science clearly clashes with neoliberalism. It is very difficult to get open data to pay for itself because open data is an almost perfect example of a public good, a type of resource markets almost invariably fail at supplying. And yet, despite public policy interest in open research data, nobody seems to know how to finance it, even at the level of the White House.[9]

While a free and open research data commons can indeed spark entrepreneurial commercial development, we enter dangerous territory by limiting our arguments to such narrow instrumentalism. Some forms of research data may have very little direct commercial interest, and may be valuable only when understood in an appropriate context. Unfortunately, neoliberal ideologies and policy making have very little time for contextualizing knowledge and knowledge creation. The *ne plus ultra* example of a neoliberal metric is the final financial return on an investment.

Let me given an example of why this hurts the cause of open science. Take a resource like the Sloan Digital Sky survey.[10] Though it lacks clear commercial potential, at least in the short term, it represents an invaluable resource for exploring basic questions in astronomy and cosmology. Such basic research, through many twists and turns, may lead to applied science and

---

[9] Federal agencies supporting research are not likely to receive additional funding to support open data services, despite the Office of Science and Technology Policy memorandum calling for open data dissemination of federally funded research (Holdren 2013).

[10] http://www.sdss.org/

engineering that can see eventual commercialization. But even more importantly, the basic research activity itself has an (admittedly diffuse and hard to measure) intrinsic value. It provides a fertile domain of fascinating questions that sharpen minds, promote analytical thinking, and spark curiosity and wonder at the world. Research in other "useless" fields like archaeology or the humanities and social sciences has a similar intrinsic value. Unfortunately, activities and outcomes that are difficult to quantify or involve wide and diffuse externalities struggle to gain recognition in neoliberal settings.

For open science to really succeed, reform advocacy needs to dismantle a powerful and entrenched set of neoliberal ideologies and policies. Some of the key benefits of open science center on diffuse and hard-to-quantify externalities, namely trust and collaboration. Trust and collaboration are key enablers in any social enterprise, including research. We erode trust at our own peril, and making up for a loss of trust through more intrusive surveillance (or metrics) exacerbates costs and dysfunctions. If we want open science to truly succeed, we need, first and foremost, to establish institutional and policy frameworks that are humane and help to cultivate community.

Conclusions

Most of this paper has focused on the underlying policy and ideological challenges that make open science difficult to institutionalize. Tinkering at the edges of a fundamentally flawed and abusive research system will do little to promote meaningful reform. Real change will require a policy and ideological commitment to making the research process more humane—not simply more productive or high-impact. That change will only come through renewed public support and financing for basic research so that competitive pressures do not kill collegiality. Meaningful reform will also require a renewed commitment to basic notions of academic freedom and autonomy so that metrics and altmetrics serve researchers, and not the other way around.

References

Albro, R 2012 *"Culture" in the Science Fictional Universe of "Big Data."* Available at http://www.ethnography.com/2012/06/culture-in-the-science-fictional-universe-of-big-data/ [Last accessed 9 April 2014].

Atici, L, Kansa, S, Lev-Tov, J, and Kansa, E 2013 Other people's data: a demonstration of the imperative of publishing primary data. *Journal of Archaeological Method and Theory* 20–4: 663–681. DOI: http://dx.doi.org/10.1007/s10816-012-9132-9. Open Access preprint available at http://escholarship.org/uc/item/1nt1v9n2.

Costello, P 2009 Motivating Online Publication of Data. *Bioscience* 59: 418–427. Available at http://dx.doi.org/10.1525/bio.2009.59.5.9 [Last accessed 29 August 2014].

Curry, S 2012 Sick of impact factors. *Reciprocal Space*. 13 August 2012. Available at http://occamstypewriter.org/scurry/2012/08/13/sick-of-impact-factors/ [Last accessed 5 August 2014].

Davidson, C 2013 *The Tragedies of Scholarly Publishing In 2013*. HASTAC. Humanities, Arts, Science and Technology Advanced Collaboratory. Available at http://hastac.org/blogs/cathy-davidson/2013/01/15/tragedies-scholarly-publishing-2013 [Last accessed 22 April 2013].

Denhart, C 2013 How The $1.2 Trillion College Debt Crisis Is Crippling Students, Parents And The Economy. *Forbes*, August 7. Accessed March 15, 2014. http://www.forbes.com/sites/specialfeatures/2013/08/07/how-the-college-debt-is-crippling-students-parents-and-the-economy/.

Fang F C and Casadevall A 2011 Retracted Science and the Retraction Index. *Infect. Immun.* 79(10): 3855–3859. Available at: http://dx.doi.org/10.1128/IAI.05661-11.

Field L 2013 *Impact sometimes requires long decades of research.* (consulted May 2014: http://www.theaustralian.com.au/higher-education/impact-sometimes-requires-long-decades-of-research/story-e6frgcjx-1226652450543).

Feller, I 2008 *Neoliberalism, Performance Measurement, and the Governance of American Academic Science.* University of California, Berkeley: Center for Studies in Higher Education (CSHE). Available at http://cshe.berkeley.edu/neoliberalism-performance-measurement-and-governance-american-academic-science [Last accessed 6 April 2014].

Garfield, E 2005 *The Agony and the Ecstasy – The History and Meaning of the Journal Impact Factor.* Available at http://garfield.library.upenn.edu/papers/jifchicago2005.pdf [Last accessed 9 April 2014].

Hamilakis, Y 2004 Archaeology and the politics of pedagogy. *World Archaeology* 36(2): 287–309. Available at https://www.academia.edu/250356/Hamilakis_Y._2004_Archaeology_and_the_politics_of_pedagogy._World_Archaeology_Vol._36_2_287-309 [Last accessed 5 August 2014].

Harley, D Krzys Acord, S, Earl-Novell, S, Lawrence, S and Judson King, C 2010 *Assessing the Future Landscape of Scholarly Communication: An Exploration of Faculty Values and Needs in Seven Disciplines.* University of California, Berkeley: CSHE. Available at http://escholarship.org/uc/item/15x7385g [Last accessed 7 October 2010].

Holdren, J P 2013 *Increasing Access to the Results of Federally Funded Scientific Research.* Memorandum from the Executive Office of the President, Office of Science and Technology Policy. 22 February. Available at http://www.whitehouse.gov/sites/default/files/microsites/ostp/ostp_public_access_memo_2013.pdf [Last accessed 15 August 2014].

Kansa E 2012 Openness and archaeology's information ecosystem. *World Archaeology* 44(4): 498–520. Preprint available at: http://escholarship.org/uc/item/4bt04063.

Kansa, E C, Whitcher Kansa, S and Arbuckle, B 2014 Publishing and pushing: mixing models for communicating research data in archaeology. *International Journal of Digital Curation* 9(1): 1–1. DOI: http://dx.doi.org/10.2218/ijdc.v9i1.301.

Kansa, E C and Whitcher Kansa, S 2013 We all know that a 14 is a sheep: data publication and professionalism in archaeological communication. *Journal of Eastern Mediterranean Archaeology and Heritage Studies* 1(1): 88–97. Open access preprint available at: http://escholarship.org/uc/item/9m48q1.

Kingfisher, C and Maskovsky, J 2008 Introduction: the limits of neoliberalism. *Critique of Anthropology* 28(2): 115–126. DOI: 10.1177/0308275X08090544.

Kratz J and Strasser C 2014 Data publication consensus and controversies [v2; ref status: indexed, http://f1000r.es/3hi] *F1000Research*, 3: 94. DOI: http://dx.doi.org/10.12688/f1000 research.3979.2.

Leetaru, K 2011 Culturomics 2.0: forecasting large-scale human behavior using global news media tone in time and space. *First Monday* 16(9). DOI: http://dx.doi.org/10.5210/fm.v16i9.3663. Available at http://firstmonday.org/ojs/index.php/fm/article/view/3663 [Last accessed 9 April 2014].

Lorenz, C 2012 If you're so smart, why are you under surveillance? Universities, neoliberalism, and new public management. *Critical Inquiry* 38(3): 599–629. DOI: http://dx.doi.org/10.1086/664553.

McCormick, T 2014 #allmetrics [Twitter] 30 January. Available at http://www.twitter.com [Last accessed 5 August 2014].

McGuigan, G S and Russell, R D 2008 *The Business of Academic Publishing*. Available at http://southernlibrarianship.icaap.org/content/v09n03/mcguigan_g01.html [Last accessed 6 April 2012].

Michel, J-B, Shen, Y K, Presser Aiden, A, Veres, A, Gray, M K, The Google Books Team, Pickett, J P, Hoiberg, D, Clancy, D, Norvig, P, Orwant, J, Pinker S and Nowak, M A 2011 Quantitative analysis of culture using millions of digitized books. *Science* 331(6014): 176–182. DOI: http://dx.doi.org/10.1126/science.1199644.

Mobbiot, G 2011 Academic publishers make Murdoch look like a socialist. *The Guardian*, 29 August. Available at http://www.guardian.co.uk/commentisfree/2011/aug/29/academic-publishers-murdoch-socialist [Last accessed 4 April 2012].

Patton, S 2012 The Ph.D. now comes with food stamps. *The Chronicle of Higher Education,* 6 May. Available at http://chronicle.com/article/From-Graduate-School-to/131795/ [Last accessed 4 April 2014].

Steele, C 2008 Scholarly Monograph publishing in the 21st century: the future more than ever should be an open book. *Journal of Electronic Publishing* 11(2). DOI: http://dx.doi.org/10.3998/3336451.0011.201.

Data Sharing in a Humanitarian Organization: The Experience of Médecins Sans Frontières

Unni Karunakara

Médecins Sans Frontières, Geneva, Switzerland

Summary Points

- Public health crises such as the spread of drug-resistant tuberculosis highlight the need for improved sharing of data. For humanitarian organizations, there is a lack of guidance on the practical aspects of making such data available.
- In 2012 the medical humanitarian organization Médecins Sans Frontières (MSF) decided to adopt a data sharing policy for routinely collected clinical and research data. Here we describe how this policy was

---

How to cite this book chapter:
Karunakara, U. 2014. Data Sharing in a Humanitarian Organization: The Experience of Médecins Sans Frontières. In: Moore, S. A. (ed.) *Issues in Open Research Data*. Pp. 59–76. London: Ubiquity Press. DOI: http://dx.doi.org/10.5334/ban.d

developed, the principles underlying it, and the practical measures taken to facilitate data sharing.
- The MSF policy builds on the principles of ethical, equitable, and efficient data sharing to include aspects relevant for an international humanitarian organization, in particular concerning highly sensitive data (non-maleficence), benefit sharing (social benefit), and intellectual property (open access).
- There are aspirations to create a truly open dataset, but the initial aim is to enable data sharing via a managed access procedure so that security, legal, and ethical concerns can be addressed.

Introduction

Open data and data sharing are essential for maximizing the benefits that can be obtained from institutional and research datasets (Murray-Rust et al. 2014). In 2012, the medical humanitarian organization Médecins Sans Frontières (MSF) decided to adopt a data sharing policy for routinely collected clinical and research data (http://www.msf.org.uk/msf-data-sharing). Here we describe the policy's principles, practicalities, and development process. We hope this paper will encourage and help other humanitarian and non-governmental organizations to share their data with public health researchers for the benefit of the populations with which they work.

The Growth of Open Data

Initiatives to promote the sharing of data generated by research activities have been led by foundations such as the Wellcome Trust and other signatories to the Full Joint Statement by Funders of Health Research (Wellcome Trust 2013), the creation of large open databases such as Dryad (2013), and journal and publisher initiatives (PLOS 2013; BioMed Central 2010; Hrynaszkiewicz 2010; Nature Publishing Group 2010). However, practical and systemic limitations have limited real data sharing across medical and clinical research (Savage and Vickers 2009) and routinely collected clinical data (Godlee 2012). Although much discussion has taken place around data sharing (Theodora Bloom, personal communication), concrete actions and a positive willingness to share data have been less common.

Datasets Collected in Humanitarian Situations

Public health crises, such as the spread of drug-resistant tuberculosis (Nyang'wa et al. 2013) and the 2002 severe acute respiratory syndrome (SARS) outbreak (World Health Organisation 2003), highlight the need for sharing data; a case has been made that data sharing is an ethical duty in such contexts (Langat et al. 2011). For humanitarian organizations, there is a lack of guidance on how and what sort of data can and should be shared, and especially on the practical aspects of making such data available while considering the sensitivities involved in datasets collected in contexts of humanitarian action.

MSF and Data Sharing

MSF and Epicentre, its research affiliate (http://www.epicentre.msf.org/en), place a high value on monitoring and documenting MSF's medical interventions to improve their quality, resulting in a large amount of routinely collected data. In addition, MSF conducts a substantial amount of operational research with patient groups and diseases commonly neglected in international research agendas (Zachariah et al. 2010; Brown et al. 2008). MSF recognizes its responsibility to share and disseminate this knowledge.

As a first step in meeting this responsibility, MSF established an institutional repository for its research publications (http://fieldresearch.msf.org/msf/) in 2008, and more recently has introduced a scientific publication policy that prioritizes open access, and is working on a policy for online sharing of research protocols.

Development of the MSF Data Sharing Policy

Until 2012, decisions to share MSF data were made on a case-by-case basis on request. Recognizing the problems inherent in this informal approach, MSF developed a proactive data sharing policy in the hope of boosting data sharing while ensuring that ethical and legal obligations were met (**Box 1**). The principles in the Full Joint Statement by Funders of Health Research (Wellcome Trust 2013) were the starting point for the MSF policy, namely, that data should be shared in a manner that is ethical, equitable, and efficient. MSF consulted with the Wellcome Trust and the MSF Ethics Review Board (Schopper et al. 2009) to adapt and expand these principles to include ones specific for MSF concerning highly sensitive data, benefit sharing, and intellectual property. The policy was drafted using a template from the UK National Cancer Research Institute (Chapman et al. 2013).

> The independent MSF Ethics Review Board was created to ensure that ethical oversight is available for issues that could arise from a humanitarian organization providing care and also requesting participation in research. In determining the procedures for our data sharing policy, two situations were identified as needing ethical review.

(Box continued on next page)

(Box continued from previous page)

> One was the inclusion of personal (identifiable) data and/or human samples (with adequate consent), given the high sensitivity of MSF contexts and—generally speaking—of human samples. Sharing of personal data or human samples potentially entails risk in terms of the perception by MSF patients and authorities in countries of operation that MSF is carrying out research under the guise of medical care. It was decided not to exclude outright the secondary use of personal (identifiable) data and/or human samples—as some of these data can be of considerable value to research that promotes health benefits. Where personal data are included in a dataset, ethical review is required.
>
> The second situation was the use of nonidentifiable research data outside of original consent agreements, which some MSF Ethics Review Board members felt should not be authorized. However, there will be rare cases of research data collected prior to the data sharing policy being created that have significant value for communities, particularly those relating to neglected diseases, where a case can be made that the benefits of sharing such data outweigh the potential harms. After considerable debate, the use of nonidentifiable research data outside of original consent agreements was accepted if MSF tries to return to study participants to expand their original consent or, failing that, is able to secure consent from the community where the study took place. Use of data outside of original consent will always require ethical review.

Box 1: Issues Requiring Ethical Review.

Vision and Principles

MSF commits to share and disseminate health data from its programs and research in an open, timely, and transparent manner in order to promote health benefits for populations while respecting ethical and legal obligations towards patients, research participants, and their communities. MSF will work towards maximizing the availability of health data of wider interest to public health researchers with as few restrictions as possible, while respecting the principles outlined in **Box 2**. Practically, these ambitions will be achieved by creating an online data collection.

> **Ethics:** MSF data sharing will abide by the following ethical principles:
>
> - **Medical confidentiality** is fully respected.
> - The **privacy and dignity of individuals** and communities are not jeopardized.
> - **Collaborative partnerships** are undertaken in line with MSF's Ethical Framework for Medical Research; recipients of MSF datasets will engage, wherever possible, with the local research community and the local community where the MSF dataset originates.
>
> **Equity:** MSF data sharing will recognize and balance the needs of practitioners or researchers who generate and use health data, other analysts who may want to reuse such data, and communities and funders who expect health benefits to arise from research.

(Box continued on next page)

(Box continued from previous page)

> **Efficiency:** MSF data sharing will improve the quality and value of the delivery of health care, and increase its contribution to improving public health. Approaches should be proportionate and build on existing practice and reduce unnecessary duplication and competition.
>
> **Non-maleficence:** Data sharing shall not put at risk, or be used against, the interests of MSF patients, MSF research participants, MSF employees, or MSF organizations for political reasons, financial gain, or any other reasons.
>
> **Social benefit:** First, to promote health benefits to the greater population, data sharing should bring health benefits to individuals and communities outside of those in which the data were collected. Second, to prioritize local benefit sharing, data sharing will prioritize data of benefit to the local communities where the data were collected, as well as to patients and communities similar to those in which MSF works, in particular marginalized or neglected populations. Notwithstanding this, there is a recognition that benefit sharing can be with a wider community of individuals, and will not always result in benefits to the local community.
>
> **Open access:** Recipients of MSF datasets shall strive to avoid prohibitively costly approaches, restrictive intellectual property strategies, or other approaches that may inhibit or delay the use of the results of their research to the benefit of low- and middle-income countries. In particular,

(Box continued on next page)

(Box continued from previous page)

> they shall put forth their best efforts to avoid anything that could seriously limit follow-up research and/or development and/or equitable and affordable access to potential final product(s) by end users in such countries. Recipients shall not seek any intellectual property rights of any kind with respect to results generated by or arising out of the use of MSF datasets without prior written consent.

Box 2: Principles Underlying Data Sharing in MSF.

Principles Developed for the MSF Data Sharing Policy

Non-maleficence

MSF projects are often located where there is political or ethnic violence, or where certain disease diagnoses are associated with government restrictions or potentially dangerous consequences. The overriding imperative for MSF is to ensure that patients are not harmed or compromised. Thus, caution is needed when handling potentially sensitive data. Sensitive data are defined as any subset of information that can be misused against the interests of the individuals whose data are included in the dataset or against MSF, or that put either individuals or MSF at risk for political, financial, or other reasons (**Box 3**). In determining the eligibility of datasets for sharing, MSF must consider their potential sensitivity and ensure that appropriate safeguards are in place. Should safeguards not be appropriate or sufficient, MSF may decide that datasets are not be eligible for sharing.

Data considered sensitive by MSF:

1. Any data from which an implication of criminal conduct could be drawn and/or that can put MSF patients or research participants at serious risk (including death). This includes data on violence-related medical activities, particularly, but not exclusively, in contexts of conflicts: (1) any data related to violence—such as bullet wounds—and (2) any data related to sexual violence.
2. Data collected from MSF activities in prisons or any situation that are related to or can result in detention or deprivation of liberty (including in certain refugee or displaced person settings).
3. Certain data variables such as those that could indirectly imply, truly or not, racial or ethnic origin, or political or religious opinions (for example, the origin or the location of the patient/participant).
4. Data related to sicknesses with an obligation to adhere to treatment.

Data considered potentially sensitive by MSF (non-exhaustive):

1. Data that can put patients/participants at risk of stigma, discrimination, or criminal sanction (including, in certain countries or populations, HIV and tuberculosis data).
2. Data on sicknesses or epidemic outbreaks.

Box 3: Sensitive Data.

Social benefit

MSF will prioritize data sharing requests that are of benefit to the local communities where the data were collected, as well as to patients and communities similar to those in which MSF works, in particular marginalized or neglected populations. Notwithstanding this, there is a recognition that benefit sharing can be with a wider community of individuals, and will not always result in benefits to the local community.

Open access

In 1999, MSF launched the Access Campaign to push for access to, and the development of, medicines, diagnostic tests, and vaccines for patients in MSF programs and beyond. Research developed as a result of data shared by MSF should remain consistent with such aims, with results and end products being accessible (and affordable) in low- and middle-income countries. In light of the potential public health benefits of releasing results immediately and without restrictions, publication of results should be consistent with the MSF scientific publishing policy, which prioritizes open access.

Access to MSF datasets will be granted only if the recipients of data agree not to seek intellectual property rights of any kind, without MSF giving specific and prior consent. In addition, recipients must avoid actions that render the results of their research, such as publications or medical products, unavailable or unaffordable for the populations of low- and middle-income countries.

What Data Will Be Included in the Data Collection?

The policy applies to all health data generated in MSF programs or sites, where MSF acts as a custodian for such data. It includes data generated from health information systems, patient records, surveillance activities, quality control activities, surveys, research, and patients' or research participants' human biological material. While the scope of the policy is purposely broad, there is no ambition to share data simply for the sake of sharing. Only data whose dissemination is judged to have the potential to lead to greater health benefits for populations will be shared (**Box 2**). Practically, this decision-making process will be implemented through a procedure whereby MSF data judged to have a substantial public health benefit are eligible to be proposed by any MSF or Epicentre staff for inclusion in the online collection. The decision to include data will be guided by the vision and principles of the data sharing policy, and data should not be unreasonably withheld. Approval for data sharing may have to be sought from other involved partners where preexisting contracts or memorandums of understanding limit data sharing.

Data initially proposed for inclusion include records of HIV treatment and care, treatment for drug-resistant tuberculosis and human African trypanosomiasis, and a database of nutritional surveys. Research data will be added as they become available.

Managed Access Procedure

Who can access the data collection?

Access to the data collection will be open to all appropriately qualified researchers from academia, charitable organizations, and private companies, such as drug companies. MSF defines an appropriately qualified researcher as someone who has authored relevant peer-reviewed articles, and who is still working in the relevant specialty (Wellcome Trust Sanger Institute 2010). We will positively consider all applications from researchers from countries and communities in which we work and, in particular, from where the specific datasets requested originated.

How will access be managed?

We intend to post some datasets in an open repository, but as a first step to gain experience with data sharing, managed access will be the default means of sharing data. A high proportion of data generated by MSF is considered sensitive, thereby requiring a higher level of oversight. The stringency of the managed access procedure will be proportionate to the risks associated with MSF datasets, and must not unduly restrict or delay access.

Costs

Most of MSF's funding comes from individual private donors who wish to support medical humanitarian assistance. Thus, MSF has chosen to implement data sharing as a cost-neutral exercise. Recipients of data will be required to cover the costs of retrieving, processing, and dispatching MSF datasets. If applicants for data sharing do not have sufficient financial means to cover such fees, exceptions can be made.

Challenges

Data Collection and Protection

The MSF data sharing policy is based on MSF's organizational commitment to improving the ethical collection and protection of data in our programs. The nature of humanitarian contexts can make this challenging, particularly in terms of the ability to obtain informed consent for data collection. Ensuring the privacy and confidentiality of the data collected also requires specific attention. For example, tissue samples have specific ethical issues attached to their collection, use, and dissemination. In MSF, material transfer agreements are now signed with external laboratories that provide advanced testing for our patients. This ensures that samples are not used without consent for purposes other than those requested by MSF clinicians, and that they are disposed of correctly.

Ensuring MSF Staff Share Data

The data sharing policy is aspirational and will rely on political engagement to ensure compliance. This is challenging because the scope of the policy with regards to routinely collected data means that the participation of MSF staff in program and headquarter offices is required, as well as that of staff involved in research, who may already appreciate the value of sharing research-generated datasets. Data sharing will be facilitated with standard templates to support development of data sharing plans and proposals.

Ensuring Inclusion of Data Sharing in Research Proposals

At the research proposal stage, if the research is likely to generate data outputs valuable for the wider public health community, MSF researchers should develop a data management and sharing plan that includes consideration of the resources required. The inclusion of a broad consent in research proposals will be considered where there is evidence of a clear potential for the greater public good and if risks are limited. Broad consent is usually granted ethics approval under the conditions that personal information is handled safely and that the donors of biological samples are granted the right to withdraw consent.

Data Quality

The value of the data sharing policy will rely on good practices in data collection, use, and management (UK Data Archive 2012). As an organization focused on providing emergency assistance, creating and maintaining datasets to a high standard is a continual challenge. Organizationally, there is commitment to strengthening standards and an expectation that data sharing itself will strengthen this process with a consistent and positive engagement with researchers and dataset managers. In addition, MSF will prioritize information technology solutions that facilitate data sharing.

Data Preservation

Preserving and protecting data from corruption or obsolescence of software is a serious concern with open data and data sharing. Digital Science offers a research data archiving service via Figshare and notes the safeguards needed to ensure the preservation and security of data (Hahnel 2012). As the MSF data sharing database grows, data preservation may require innovative thinking to ensure its security.

The Way Forward

MSF's core mission is to respond to medical humanitarian crises. This priority makes it quite unlike the large research-oriented organizations and funders that have pioneered data sharing. MSF's data sharing policy will test the ability of the organization to protect the vulnerable population it serves while contributing to health research to ultimately benefit the communities and patients from which the data were gathered.

Acknowledgments

This article is authored by Unni Karunakara on behalf of Médecins Sans Frontiéres. The MSF Data Sharing Working Group developed the MSF data sharing policy. The members of the group were Unni Karunakara, International President of MSF; Emmanuel Baron, Executive Director of Epicentre; Ondine Ripka, MSF Legal Department; and Leslie Shanks, Medical Director, Operational Centre Amsterdam.

We thank Doris Schopper and Ross Upshur of the MSF Ethics Review Board for helpful comments on the manuscript, and Caley Montgomery for editing assistance.

Sarah Venis wrote the first draft of the paper (and reviewed and edited later revisions) based on literature review, interview of an expert in open data management (Theodora Bloom), and the data sharing policy.

Editor's note

This article originally appeared in the journal *PLOS Medicine* and is reproduced in accordance with the CC BY licence and with kind permission of Dr. Karunakara. Whilst there have been no alterations to the content, the reference style has been amended for consistency with the other chapters in the book. The original citation for the article is:

Karunakara U (2013) Data Sharing in a Humanitarian Organization: The Experience of Médecins Sans Frontières. PLoS Med 10(12): e1001562. doi:10.1371/journal.pmed.1001562.

References

BioMed Central (2010) *BioMed Central's position statement on open data.* Available at: http://blogs.biomedcentral.com/bmcblog/files/2010/09/opendatastatementdraft.pdf [Accessed 16 April 2013].

Brown, V., Guerin, P., Legros, D., Paquet, C., Pécoul, B. and Moren, A. (2008). Research in Complex Humanitarian Emergencies: The Médecins Sans Frontières/Epicentre Experience. *Plos Med*, [online] 5(4), p.e89. Available at: http://dx.doi.org/10.1371/journal.pmed.0050089 [Accessed 16 Sep. 2014].

Chapman, M., Carrigan, C., Clark, B., Cope, J. and Groot, K. (2013). *A template for the development of policies for access to data or biological samples for research.* [online] London: National Cancer Research Institute. Available at: http://www.ncin.org.uk/view?rid=250 [Accessed 4 Nov. 2013].

Dryad (2013) Dryad [data repository]. Available at: http://datadryad.org/ [Accessed 16 April 2013].

Godlee, F. (2012). Measure your team's performance, and publish the results. *BMJ*, [online] 345(jul04 2), pp.e4590–e4590. Available at: http://dx.doi.org/10.1136/bmj.e4590 [Accessed 16 Apr. 2013].

Hahnel M (2012) *Ensuring persistence on figshare.* Available at: http://figshare.com/blog/Ensuring%20persistence%20on%20figshare/25 [Accessed 16 April 2013].

Hrynaszkiewicz, I (2010) *A call for BMC Research Notes contributions promoting best practice in data standardization, sharing and publication.* BMC Res Notes 3: 235 Available at: http://www.biomedcentral.com/1756-0500/3/235/ [Accessed 16 April 2013].

Langat, P., Pisartchik, D., Silva, D., Bernard, C., Olsen, K., Smith, M., Sahni, S. and Upshur, R. (2011). Is There a Duty to Share? Ethics of Sharing Research Data in the Context of Public Health Emergencies. *Public Health Ethics*, [online] 4(1), pp.4–11. Available at: http://dx.doi.org/10.1093/phe/phr005 [Accessed 9 Sep. 2013].

Murray-Rust, P., Neylon, C., Pollock, R. and Wilbanks, J. (2014). *Panton Principles*. [online] Pantonprinciples.org. Available at: http://pantonprinciples.org [Accessed 16 Apr. 2013].

Nature Publishing Group (2012) *Availability of data and materials*. Available: http://www.nature.com/authors/policies/availability.html [Accessed 16 April 2013].

Nyang'wa, B., Brigden, G., du Cros, P. and Shanks, L. (2013). Resistance to second-line drugs in multidrug-resistant tuberculosis. *The Lancet*, [online] 381(9867), p.625. Available at: http://dx.doi.org/10.1016/s0140-6736(13)60341-4 [Accessed 16 Apr. 2013].

PLOS (2013) *PLOS editorial and publishing policies*. Available at: http://www.plosone.org/static/policies.action#sharing [Accessed 16 April 2013].

Savage, C. and Vickers, A. (2009). Empirical Study of Data Sharing by Authors Publishing in PLoS Journals. *PLOS ONE*, [online] 4(9), p.e7078. Available at: http://dx.doi.org/10.1371/journal.pone.0007078 [Accessed 16 April. 201].

World Health Organisation, (2003). *Consensus document on the epidemiology of severe acute respiratory syndrome (SARS)*. Geneva: Consensus document on the epidemiology of severe acute respiratory syndrome (SARS). http://apps.who.int/iris/bitstream/10665/70863/1/WHO_CDS_CSR_GAR_2003.11_eng.pdf [Accessed 9 September 2013].

Schopper, D., Upshur, R., Matthys, F., Singh, J., Bandewar, S., Ahmad, A. and van Dongen, E. (2009). Research Ethics Review in Humanitarian Contexts: The Experience of the Independent Ethics Review Board of Médecins Sans Frontières. *PLoS Med*, [online] 6(7), p.e1000115. Available at: http://dx.doi.org/10.1371/journal.pmed.1000115.

UK Data Archive (2012) *Create and manage data: planning for sharing. How to share data.* Available at: http://www.data-archive.ac.uk/create-manage/planning-for-sharing/how-to-share-data [Accessed 16 April 2013].

Wellcome Trust Sanger Institute (2010) *Data sharing guidelines.* Available at: http://www.sanger.ac.uk/datasharing/assets/wtsi_datasharing_guidelines.pdf [Accessed 16 April 2013].

Wellcome Trust, (2013). *Sharing research data to improve public health: full joint statement by funders of health research | Wellcome Trust.* [online] Available at: http://www.wellcome.ac.uk/About-us/Policy/Spotlight-issues/Data-sharing/Public-health-and-epidemiology/WTDV030690.htm [Accessed 16 Apr. 2013].

Zachariah, R., Ford, N., Draguez, B., Yun, O. and Reid, T. (2010). Conducting operational research within a non governmental organization: the example of Médecins Sans Frontières. *International Health*, [online] 2(1), pp.1–8. Available at: http://dx.doi.org/10.1016/j.inhe.2009.12.008.

Why Open Drug Discovery Needs Four Simple Rules for Licensing Data and Models

Antony J. Williams,[*] John Wilbanks[†] and Sean Ekins[‡]

[*]Royal Society of Chemistry, Wake Forest, North Carolina, USA
[†]Consent to Research, Oakland, California, USA
[‡]Collaborations in Chemistry, Fuquay-Varina, North Carolina, USA

Introduction

Public online databases supporting life sciences research have become valuable resources for researchers depending on data for use in cheminformatics, bioinformatics, systems biology, translational medicine, and drug repositioning efforts, to name just a few of the potential end user groups (Williams et al. 2009). Worldwide funding agencies (governments and not-for-profits) have invested in public domain chemistry platforms. In the United States these include PubChem, ChemIDPlus, and the Environmental Protection Agency's ACToR, while the United Kingdom

---

How to cite this book chapter:
Williams, A. J., Wilbanks, J. and Ekins, S. 2014. Why Open Drug Discovery Needs Four Simple Rules for Licensing Data and Models. In: Moore, S. A. (ed.) *Issues in Open Research Data*. Pp. 77–88. London: Ubiquity Press. DOI: http://dx.doi.org/10.5334/ban.e

has funded ChEMBL and ChemSpider, among others, and new databases continue to appear annually (National Center for Biotechnology Information, n.d.; US National Library of Medicine, n.d.; Judson et al. 2008; EMBL-EBI, n.d.; Pence & Williams 2010; Galperin & Cochrane 2011).

We have argued recently that the data quality contained within many of these databases is suspect and scientists should consider issues of data quality when using these resources (Williams et al. 2011a; 2012a). By assimilating various data sources together and meshing data on drugs, proteins, and diseases, these various databases and network and computational methods may be useful to accelerate drug discovery efforts. The development of related cheminformatics platforms or derived models without care given to data quality is a poor strategy for long-term science as errors become perpetuated in additional databases (Fourches et al 2010). There is real evidence that the integration of large, heterogeneous sets of databases and other types of content is "unreasonably effective" at accelerating the conversion of data into knowledge (Halevy et al. 2009). This implies the need for technical and semantic work to bring databases together that were never designed for interoperability, which is in itself a significant task (Sansone et al. 2012; NeuroCommons, n.d.; Ruttenberg et al. 2009).

As we and others have argued previously, there is another dimension to interoperability than technical formats and ontological agreement: the complex interactions of database licenses and terms of use around intellectual property (Sansone et al. 2012; Hasting et al 2011). Many of these online databases have either obscure or confused licensing terms, and even in those cases where data are freely available for download and reuse there are often no clear definitions (de Rosnay 2008). Many databases simply "cut and paste" prohibitive copyright schema from traditional websites, or fail to address download and reintegration entirely (de Rosnay 2008). Since copyright law requires explicit permissions in advance to make use of copyrighted works, it is certainly unsafe to assume data licensing rights for any database that does not explicitly allow it.

The availability of data for download and reuse is an important offering to the community, as these data may be used for the purpose of modeling to develop prediction tools (Ekins & Williams 2010). In addition, data can be ingested into internal systems inside pharmaceutical companies to mesh with their existing private data, including in the expanding Linked Open Data cloud or in freely available online databases, and can be downloaded and used to enhance their content and to establish linking between data (Zhu et al. 2010). The Open PHACTS project, utilizes a semantic web approach to integrate chemistry and biology data across a myriad of data sources, including for chemistry ChEBI, ChEMBL, and DrugBank, and for biology UniProt, Wikipathways, and many others (Azzaoui 2012; Williams et al 2012b). The chemical structure representations are obtained from ChemSpider, which has previously imported the chemical databases and standardized according to their data model and are making the data available as open data to the project. Many of the primary online databases already have multiple links to external systems. This linking may be achieved by using available database services to form transitory links in by, for example, using a chemical representation such as an InChI to probe an application programming interface, search for the compound, and generate the linking URL in real time (Wikipedia, n.d.). Commonly, however, the links are more permanent in nature and are generated by downloading data from the various data sources, depositing a subset of the data (generally the chemical compound and associated database identifier), and using the particular database URL structure to form permanent links. This act of download and deposition of multiple data sources is commonly mixing the various licenses, if licenses are even declared, which, in many cases, they are not.

In some ways, there are analogous difficulties in the exchange of computational models like quantitative structure activity relationship (QSAR) datasets—while there are efforts to standardize how the data and models are stored, queried, and exchanged, there has been little consideration of licenses required to enable making the sharing of open source models a reality (Spjuth 2010; Gupta 2010). Similarly, one could consider the creation of maps of disease and how they are shared and reused [24] in the same manner (Derry et al. 2012).

The potential legal fragility of knowledge products derived from online databases with poorly understood licensing for each of the databases is a real problem, and one that will only increase in severity over time. This realization is not novel; indeed, the chemical blogosphere has been host to many discussions regarding the need for clear data licensing definitions on chemistry-related data. Many scientists likely echo these comments, but we will provide some examples. In particular, Peter Murray-Rust espouses the value of "open data" to the scientific discovery process and encourages clear licensing of all chemistry data according to Open Knowledge Definition (OKD) and the Panton Principles (Murray-Rust, n.d.; Wikipedia, n.d.(a); Open Knowledge Foundation, n.d; Murray-Rust et al. 2010).

Herein we provide an extensive background to the intellectual property around data and databases in the sciences involved in drug discovery, those of biology, chemistry, and related fields, as well as discussion of open data licensing, openness, and open license limitations (Text S1). More importantly, we provide a set of rules that practitioners might apply when making data or databases available via the Internet or mobile apps (Williams et al. 2009b). Our ultimate goal is to illuminate the legal fragility of the database ecosystem in the drug discovery sciences, and to initiate a conversation about creating best practices.

Simple Rules for Licensing "Open" Data

We suggest based on our analysis of the current data situation (Text S1) the ideal is to use strong default rules for openness. From a copyright and database rights perspective, the public domain gives the most clarity and should be the default setting for data deposit, although it may not always be achievable. Understanding this is vital, because it sets the bar at the right height. Justifications for additional controls should be subject to argument—one often finds those controls are unnecessary when the discussion is framed this way.

It is also important to avoid noncommercial or share-alike approaches whenever possible. These are attractive terms to many data providers, but create significant barriers to interoperability. Noncommercial data might be incompatible for researchers at a pharmaceutical company, even to run a simple web-based query. It is important to realize data under a share-alike license from one entity is probably not combinable with data under a share-alike license from another entity (this lack of interoperability kept Creative Commons licensed images out of Wikipedia for years, and is not one we wish to introduce into the ecosystem again!).

Thus, we propose the following simple rules for developing data licensing approaches inside scientific projects.

1. Before you begin a database project, convene a meeting of all of the stakeholders. Expose all of the expectations

of the group and decide if your goals are primarily scientific, commercial, or mixed. If mixed, take a stern look at the actual commercial potential of the project. Invite technology transfer offices to join you—they have greater experience in the realities of commercialization.
2. If your project is scientific in nature, and not commercial, explore the benefits of open licensing and drawbacks of enclosure. Go through the various definitions and find the most common ground possible, always placing the burden of proof on those who want more control and not less. This will create less "default enclosure" but allow for those increasingly rare situations in which "open" is not appropriate. Attempt to hew as closely as possible to the admittedly rigorous open definitions and standards, and do not write your own intellectual property licenses—instead, use existing and well deployed ones.
3. Develop simple explanations of your terms of use, and make them easy to find for users. Make sure that your licensing, expectations for attribution, terms of use, and more are linked in many ways to your data and database. Do not expect your users to read the legal text of your terms and conditions and licenses; instead, create simple summaries with linkages to the detailed text for users to access. Whenever possible, use metadata to indicate the licensing terms explicitly—the Creative Commons Rights Expression Language is a good tool for this (Creative Commons, n.d.).
4. Don't ever lock up metadata. A significant swath of data will be incompatible with an open regime, whether it's to protect trade secrets or patient privacy. But the metadata that describes closed data, and how to access closed

data, can be almost as valuable. If you can't make the data public domain, make the metadata public domain.

As a general rule, these four simple rules should allow us to build a more stable data and model sharing ecosystem while we live with some uncertainties until the courts rule on where the line of property stops and starts. We can't wait for the certainty to emerge, but we also want our systems to work when the courts do finally rule on issues such as where data and metadata stop and start, where copyright attaches, how data rights really affect re-use, and what it means to move towards a "cloud world" where copies aren't made of data at all. Following these heuristics when providing and/or accepting data is an approach that creates at least the opportunity to be forward-compatible for the future development of technologies.

But it is also important to pay close attention to licensing sanitation as a data consumer and user. No matter how tempting it is, do not copy a batch of informally open, but formally closed, data, run a database integration, and release the new database as "open"—that hurts the community. Instead, look for the terms of use, ask if it is "open", post your enquiry, and only when you are certain, redistribute. We think databases funded by the government should at the very least be open, and if not this should be stated prominently.

Conclusions

Although most scientists are likely unaware of this at present, data licenses are going to become increasingly important in science in the future, especially as we see more scientists embracing open notebook science, open science, and open-access publishing, and funding bodies promoting the increased accessibility of the fruits of their funding. We are likely not too far from funding bodies mandating immediate release of all data and results produced by each of their grantees, which is something we would advocate as potentially disruptive in its own right (S. Ekins et al., unpublished data).

We can hence imagine a near future in which many scientists will blog some or all of their research results while data aggregators will in turn consume this content and repackage it for others (Ekins et al. 2012). The licensing of this and other data will need to be clear if we are to build on the shoulders of giants and not have to face legal battles that pit Davids versus Goliaths. Considering data licensing as a part of the "scientific process" is vital for its future usability, and we strongly encourage scientists to consider data licensing before they embark upon re-using such content in databases they construct themselves or in the course of their research.

The four simple rules we have formulated for licensing data for open drug discovery represent a proposed starting point for consideration by database producers. These licenses could equally be used by individual scientists on their blogs and other online environments or accounts in which they make their data and models available for others.

Text S1

The supporting text file can be accessed at the following location http://doi.org/10.1371/journal.pcbi.1002706.s001 [PDF]. This consists of a discussion in three sections:

- Intellectual property rights in data: Copyright and Database Rights.
- Trends in legal certainty: Open Data Licensing.
- "Informal" Openness and Open License Limitations.

Editor's note

This article originally appeared in the journal *PLOS Computational Biology* and is reproduced in accordance with the CC BY licence and with kind permission of the authors. Whilst there have been no alterations to the content, the reference style has been amended for consistency with the other chapters in the book. The original citation for the article is:

Williams AJ, Wilbanks J, Ekins S (2012) Why Open Drug Discovery Needs Four Simple Rules for Licensing Data and Models. PLoS Comput Biol 8(9): e1002706. doi:10.1371/journal.pcbi.1002706

References

Azzaoui K, Jacoby E, Senger S, Rodríguez EC, Loza M, et al. (2012) Analysis of the scientific competency questions followed by the IMI OpenPHACTS consortium for the development of the semantic web-based molecular information system OPS. *Drug Disc Today* In press.

Creative Commons (n.d.) ccREL: Creative Commons rights expression language. Available at: http://www.w3.org/Submission/ccREL/ [Accessed August 2012].

de Rosnay MD (2008) Check your data freedom: a taxonomy to assess life science database openness. *Nature Precedings* Available at: http://dx.doi.org/10.1038/npre.2008.2083.1 [Accessed August 2012].

Derry JM, Mangravite LM, Suver C, Furia MD, Henderson D, et al. (2012) Developing predictive molecular maps of human disease through community-based modeling. *Nat Genet* 44: 127–130. doi: 10.1038/ng.1089.

Ekins S, Williams AJ (2010) Precompetitive preclinical ADME/Tox Data: set it free on the web to facilitate computational model building to assist drug development. *Lab on a Chip* 10: 13–22. doi: 10.1039/b917760b.

Ekins S, Clark AM, Williams AJ (2012) Open drug discovery teams: a chemistry mobile app for collaboration. *Molecular Informatics*. doi:10.1002/minf.201200034.

EMBL-EBI (n.d.) ChEMBL. Available: http://www.ebi.ac.uk/chembldb/index.php [Accessed August 2012].

Fourches D, Muratov E, Tropsha A (2010) Trust, but verify: on the importance of chemical structure curation in cheminformatics and QSAR modeling research. *J Chem Inf Model* 50: 1189–1204. doi: 10.1021/ci100176x.

Galperin MY, Cochrane GR (2011) The 2011 Nucleic Acids Research Database issue and the online Molecular Biology Database Collection. *Nucleic Acids Res* 39: D1–D6. doi: 10.1093/nar/gkq1243.

Gupta RR, Gifford EM, Liston T, Waller CL, Bunin B, et al. (2010) Using open source computational tools for predicting human metabolic stability and additional ADME/TOX properties. *Drug Metab Dispos* 38: 2083–2090. doi: 10.1124/dmd.110.034918.

Halevy A, Norvig P, Pereira F (2009) The unreasonable effectiveness of data. *Intelligent Systems* 24: 8–12. doi: 10.1109/mis.2009.36.

Hastings J, Chepelev L, Willighagen E, Adams N, Steinbeck C, et al. (2011) The chemical information ontology: provenance and disambiguation for chemical data on the biological semantic web. *PLoS ONE* 6: e25513. doi:10.1371/journal.pone.0025513.

Judson R, Richard A, Dix D, Houck K, Elloumi F, et al. (2008) ACToR–Aggregated Computational Toxicology Resource. *Toxicol Appl Pharmacol* 233: 7–13. doi: 10.1016/j.taap.2007.12.037.

Murray-Rust P (n.d.) Dr Peter Murray-Rust. Available at: http://www.ch.cam.ac.uk/person/pm286 [Accessed August 2012].

Murray-Rust P, Neylon C, Pollock R, Wilbanks J, Open Knowledge Foundation Working Group on Open Data in Science (2010) The Panton principles. Available at: http://pantonprinciples.org/ [Accessed August 2012].

National Center for Biotechnology Information (n.d.) The PubChem database. Available: http://pubchem.ncbi.nlm.nih.gov/ [Accessed August 2012].

NeuroCommons (n.d.) NeuroCommons project. Available at: http://neurocommons.org [Accessed August 2012].

Open Knowledge Foundation (n.d.) Open data licensing. Available at: http://wiki.okfn.org/Open_Data_Licensing [Accessed August 2012].

Pence H, Williams AJ (2010) ChemSpider: an online chemical information resource. *J Chem Educ* 87: 1123–1124. doi: 10.1021/ed100697w.

Ruttenberg A, Rees JA, Samwald M, Marshall MS (2009) Life sciences on the Semantic Web: the Neurocommons and beyond. *Brief Bioinform* 10: 193–204. doi: 10.1093/bib/bbp004.

Sansone SA, Rocca-Serra P, Field D, Maguire E, Taylor C, et al. (2012a) Toward interoperable bioscience data. *Nat Genet* 44: 121–126. doi: 10.1038/ng.1054.US National Library of Medicine (n.d.) ChemIDPlus Advanced. Available: http://chem.sis.nlm.nih.gov/chemidplus/ [Accessed August 2012].

Spjuth O, Willighagen EL, Guha R, Eklund M, Wikberg JE (2010) Towards interoperable and reproducible QSAR analyses: exchange of datasets. *J Cheminform* 2: 5. doi: 10.1186/1758-2946-2-5.

Wikipedia (n.d.) InChIKey on the InChI Wikipedia page. Available at: http://en.wikipedia.org/wiki/International_Chemical_Identifier#InChIKey [Accessed August 2012].

Wikipedia (n.d.(a)) Open data. Available at: http://en.wikipedia.org/wiki/Open_data [Accessed August 2012].

Williams AJ, Tkachenko V, Lipinski C, Tropsha A, Ekins S (2009) Free online resources enabling crowdsourced drug discovery. *Drug Discovery World* 10, Winter 33–38.

Williams AJ, Ekins S (2011a) A quality alert and call for improved curation of public chemistry databases. *Drug Disc Today* 16: 747–750. doi: 10.1016/j.drudis.2011.07.007.

Williams AJ, Ekins S, Clark AM, Jack JJ, Apodaca RL (2011b) Mobile apps for chemistry in the world of drug discovery. *Drug Disc Today* 16: 928–939. doi: 10.1016/j.drudis.2011.09.002.

Williams AJ, Ekins S, Tkachenko V (2012) Towards a gold standard: regarding quality in public domain chemistry databases and approaches to improving the situation. *Drug Discov Today* 17: 685–701. doi: 10.1016/j.drudis.2012.02.013.

Williams AJ, Harland L, Groth P, Pettifer S, Chichester C, et al. (2012b) Open PHACTS: semantic interoperability for drug discovery. *Drug Discov Today*, In press. Available: http://dx.doi.org/10.1016/j.drudis.2012.05.016 [Accessed August 2012].

Zhu Q, Lajiness MS, Ding Y, Wild DJ (2010) WENDI: a tool for finding non-obvious relationships between compounds and biological properties, genes, diseases and scholarly publications. *J Cheminform* 2: 6. doi: 10.1186/1758-2946-2-6.

Open Data in the Earth and Climate Sciences

Sarah Callaghan
British Atmospheric Data Centre, UK

Introduction

It is commonly acknowledged that data is the foundation of science—without access to the data used to derive results and conclusions it is not possible for other researchers to verify and reproduce the science. Reproducibility, though a fundamental part of the scientific process, is a difficult principle to follow for a number of reasons. This is especially true in the Earth and climate sciences, where even a simple experiment of taking an outdoor air temperature measurement may vary from one minute to the next, with no possibility of repeating measurements that occurred in the past.

How to cite this book chapter:
Callaghan, S. 2014. Open Data in the Earth and Climate Sciences. In:
 Moore, S. A. (ed.) *Issues in Open Research Data*. Pp. 89–106. London:
 Ubiquity Press. DOI: http://dx.doi.org/10.5334/ban.f

Access to and openness of data will facilitate reproducibility of science in the future. In the present, access to data encourages increased collaboration and reuse, allowing the identification of new multidisciplinary research avenues.

Along with the principle of reproducibility, openness of data in the Earth sciences allows for a better understanding of vital systems, including climate and weather. It is simply not possible for researchers to take measurements of every meteorological parameter at every point on the surface of the Earth. Past weather measurements, such as those found in historical ships logs (Oliver & Kington 1970; Garcia-Herrera et al. 2005; Chappell & Lorrey 2013), are invaluable for filling in the gaps in our understanding of climate change.

The Challenges of Earth and Climate Science Data

The majority of Earth science data is observational, which means that it is irreproducible. Without the aid of a time machine, it is simply not possible to travel back to last week to take a measurement that was forgotten at the time. For the same reason, we need to manage and archive the data that was collected last week, because if it is lost, it is gone for good. This is particularly relevant for fast-changing phenomenon such as weather, whereas the timescales for measurement are a bit more forgiving when it comes to the geological sciences (though not always—see for example the differences in measurements of Mount St Helens mere minutes before and after its eruption (US Geological Survey 2000)).

By contrast, much climate science is done using large and complicated software models to simulate the climate. In theory, because these are computer models, the results are reproducible by simply re-running the model with the same input parameters. In practice, however, this is not possible due to the complexity of the models, and a lack of standardisation of the metadata required to initiate them and reproduce model runs. The recent European Union Framework 7 project Metafor (Guilyardi et al. 2013) attempted to standardise and collect the metadata needed for the climate model runs done as part of the Fifth Climate Intercomparison Project (CMIP5), which fed into the Fifth Assessment Report of the Intergovernmental Panel on Climate Change. Metafor used a web-based questionnaire-type system, with associated controlled vocabulary, which took climate-modelling centres approximately two weeks to fill in—a not inconsiderable effort!

Earth science data also comes in a wide variety of formats (almost one for each type of measuring instrument used), and the datasets produced can get up to terabytes in size, as well as taking months, years, or even decades to complete.

As an (incomplete) example, the UK's Natural Environment Research Council (NERC) funds seven data centres that between them have responsibility for the long-term management of NERC's environmental data holdings. These data centres deal with a variety of environmental measurements, along with the results of model simulations in atmospheric science; Earth sciences; Earth observation; marine science; polar science; terrestrial and freshwater science, hydrology and bioinformatics; and solar-terrestrial physics and space weather.

The NERC environmental data centres hold many different types of datasets, including time series, with some series some being continually updated (e.g. meteorological measurements); large four-dimensional synthesised datasets (e.g. climate, oceanographic, hydrological, and numerical weather prediction model data generated on a supercomputer); two-dimensional scans (e.g. satellite data, weather radar data); two-dimensional snapshots (e.g. cloud cameras); traces through a changing medium (e.g. radiosonde launches, aircraft flights, ocean salinity and temperature); datasets consisting of data from multiple instruments as part of the same measurement campaign; and physical samples (e.g. fossils).

Data is also produced in a variety of ways by a variety of researchers, ranging from individual researchers, to small research groups, up to entire institutions.

Large research groups and institutions tend to have a more 'industrial' process for developing the data, where standards for data formats and metadata are well defined and adhered to by all participants. Openness of the data within the collaboration or project group is the norm, and systems are set in place to share the data within that group. The standardised data formats and metadata are a boon to helping the project members share data within their group, and would be useful for researchers using the data outside the group too. Often, however, the data are closed to all but the members of the group. Paradoxically, putting access restrictions in place on a collaborative workspace may make researchers more likely to open their data within that workspace and begin the process of sharing.

Small research groups are less likely to have standardised formats for data and/or metadata (unless they are part of a larger community, such as the atmospheric sciences where standardised file formats such as NetCDF are common). This does not make them any less open to sharing their data, but it does introduce an extra overhead of effort for the person being shared with, as they then have to learn the format and decipher the metadata (if any) before they can use the dataset.

Drivers for Openness

Measuring Earth science phenomena is expensive, often requiring expensive equipment such as ships or aircraft, or large networks of instruments such as rain gauges or radars. Funders are keen to ensure that the data collected as a result of their funding is archived and managed properly, not only to ensure the quality of the research, but also to enable reuse of the data by other researchers (both inside the domain of interest and related) thereby saving time, effort and money.

Members of the climate science community were pushed towards openness after the 'Climategate' affair, when, in November 2009, a server at the Climatic Research Unit (CRU) at the University of East Anglia was hacked and thousands of emails and computer files were copied to various locations on the Internet. This resulted in the spread of alleged malpractices found within the leaked CRU emails around the Internet, where it was claimed 'that these e-mails showed a deliberate and systematic attempt by leading climate scientists to manipulate climate data, arbitrarily adjusting and 'cherry-picking' data that supported their global warming claims and deleting adverse data that questioned their theories.' (House of Commons Science and Technology Committee 2010). In the resulting investigations, no evidence of fraud or scientific misconduct was found, and recommendations were made to avoid any such allegations in the future by opening up access to their supporting data.

Openness of data encourages reuse, and adds value to other research. One example is of a researcher using rainfall data in her studies of newt populations (British Atmospheric Data Centre 2007); her access to this data added an extra dimension to her studies, allowing her to draw more complete conclusions. Without access to the data, her research would have been the poorer, as she would not have been able to make the required measurements herself in the context of her own investigation.

Barriers to Openness

Simply opening up a dataset for use by others is not enough. It is very easy to stick some data files on a departmental or personal website, with file names that may be clear to the producer, but are opaque to everyone else. Once in the file (assuming they can open it in the first place), a potential user may have to figure out what the various variables actually mean, and then dig through other information (published in journal articles or not) before they can really make use of the data. Just because data is open does not mean it is usable. Similarly, just because a dataset is archived does not mean it will still be usable in 20 years' time, especially given the rate of change in commonly used file formats such as Excel.

In an increasingly competitive environment for research funding, access to important datasets may be the only factor determining whether a grant is won or not. For this reason, there is a tendency for researchers to hoard data until they have extracted all the possible research benefit out of it. This can be combated by the research funders' policies on open data and embargoes.

In the absence of common practices or standards, some researchers have misgivings about making their data either freely or openly available, as they fear that other researchers may find errors or 'misuse the data', or that the researcher themselves will get 'scooped' and lose out on research funding (RIN 2008).

A Tale of Two Datasets: The Author's Personal Experience of Open Data

The datasets

Upon finishing her first degree, the author was hired by the Radio Communications Research Unit (RCRU, now the Chilbolton Group), at Science and Technology Facilities Council Rutherford Appleton Laboratory, UK, to process and analyse data received from ITALSAT, a communications satellite. The group was investigating the effects of rain, clouds and atmospheric gases on the received signal levels from radio beacons aboard geosynchronous satellites. Their aim was to determine the best way of counteracting the signal fading experienced by radio frequencies above 10 GHz when a rain storm blocks the path between the satellite transmitter and the receiver in the ground station. To perform these measurements, the RCRU installed and operated a number of receivers at different locations in Hampshire, UK. **Table 1** gives information about the experiments, including the locations, the measurement periods and the primary publications. It is important to note the significant delay between the completion of the ITALSAT experiment and the primary publication from it. Also, the primary work of the staff involved in the ITALSAT and Global Broadcast Service (GBS) experiments was to run and manage long-term measurement campaigns, meaning that writing up the experiments for publication was often a lower priority.

Pre-processing the received signal levels was the author's main job for several years. The received signal levels had to be processed to remove the diurnal variability introduced as the satellite varied in its orbit because of its age and the lack of fuel available to make station keeping adjustments. This pre-processing

Experiment	ITALSAT	GBS (Global Broadcast Service)
Frequencies studied	49.5, 39.6 and 18.7 GHz	20.7 GHz
Receive sites	Sparsholt (51° 04' N, 01° 26' W)	Sparsholt (51° 04' N, 01° 26' W) Chilbolton (51° 08' N, 01° 26' W) Dundee (56.45811° N, 2.98053° W)
Measurement period start	April 1997	Chilbolton: August 2003 Sparsholt: October 2003 Dundee: February 2004
Measurement period end	January 2001	August 2006
Primary publication(s)	Ventouras et al. 2006	Callaghan et al. 2008 Callaghan et al. 2013

Table 1: Key characteristics of the ITALSAT and GBS datasets.

involved four major steps, four different computer programmes and 16 intermediate files for each day of measurements. Each month of pre-processed data represented somewhere between a couple of days' and a week's worth of effort. It was a job where attention to detail and scientific knowledge and data experience were important.

Sharing the data

The ITALSAT raw and processed data were stored on the RCRU's servers, with a backup on CD on a shelf in the author's office (where it still resides).

We were approached by other radio propagation research groups to share our data, and in some cases we did so. Because the data were in a non-standard format, this involved sharing the software we used and, occasionally, physically sitting with the new users, explaining how it had been created and what the files meant.

The first article about the ITALSAT dataset was published in 2003 (Otung & Savvaris 2003), three years before the first publication from the researchers who produced the data. We were not part of the author list on the 2003 paper, though I believe we got a group acknowledgement.[1] There was also at least one other occasion where we 'shared' our data with other researchers, who then went on to receive further funding for work in the same subject area that did not include us.

An added complication was that this data was (in theory) commercially valuable and could have been sold to telecommunications companies. Hence, in a number of cases, sharing required the development of non-disclosure agreements, in consultation with our contracts department, which took a lot of time and effort.

Eventually, we just hoarded the data, which was not good for us, or for science! It was only after the group's funding was changed, and our new funders mandated that all the group's data should be deposited in the British Atmospheric Data Centre (BADC), that we moved away from keeping the data on private servers in non-standard formats.

Opening the data

Both the ITALSAT and GBS datasets have now been archived in the BADC and have been assigned digital object identifiers (DOIs) to enable formal data citation to occur (STFC 2009a, 2009b, 2009c, 2012a, 2012b, 2012c). It is worth noting that the

---

[1] Unfortunately I cannot check as the referenced paper is behind a paywall.

DOIs for the GBS dataset were only assigned in April 2011, and the ITALSAT data DOIs were assigned in 2012—a long time after the completion of the datasets and their primary publications. Even though the datasets are now citeable and discoverable in the BADC, they are still not completely open, as they can only be downloaded by registered BADC users. However, there are no restrictions on who can become a BADC user. Also, the Chilbolton Group would like to monitor the use of these data and require an acknowledgement of the data source if they are used in any publication.

Detailed project reports were written about both the ITALSAT and GBS experiments and provided to the funders of the experiments. These reports are a valuable resource because they are significantly longer and more detailed than the journal publications, but because they are grey literature, access to them is limited. For the GBS experiment, the report is marked as 'commercial in confidence' and therefore cannot be made public. For ITALSAT, the documentation has fallen foul of changes in word processing software and key figures in the document cannot be viewed on-screen. This just goes to show that data curation applies to supporting documentation as much as it does to the datasets themselves.

Publications and the datasets

Ventouras and colleagues (2006) do not make any statement on data availability or the location of the raw data. The article does include some of the derived data in the form of tables and figures of cumulative distribution functions, but there is a crucial disconnect between the paper and the dataset on which it bases its conclusions.

Similarly, for the GBS dataset, the Callaghan et al. (2008) paper does not include any figures or tables of the processed data, instead only presenting figures showing the curves resulting from the analysis. These authors do comment about the location of the underlying data: 'The database collected as part of the GBS experiment has been submitted to the International Telecommunications Union (ITU-R) Study Group 3 for inclusion into its databanks.' These databanks are available online but it is not clear where the GBS experiment data can be found within them.[2]

Note that for both experiments, the final step (archiving the data or publication in a data journal) took place some time after the experiment was officially concluded. This would not be possible for many research groups because the researcher who did the majority of the data processing and analysis is very likely to have left that research group (as a result of finishing their PhD or postdoc, or finding a position elsewhere once the project funding finished).

Encouraging Openness: Carrots and Sticks

As mentioned earlier, the scientific consensus is changing to the belief that openness should be the norm rather than the exception (Royal Society 2012). But in order to encourage the researcher producing the data to open it and, more to the point, open it in a way that is useful to other users, rewards and sanctions are needed. Steps have already been made, with many research funders publishing data policies (RCUK 2013a, b; European Commission 2013; NSF 2010) that outline their expectations of their funded

---

[2] http://www.itu.int/ITU-R/index.asp?category=study-groups&rlink=rsg3&lang=en

researchers. The methods for applying sanctions have yet to be applied, or even defined.

Focus in the UK and elsewhere has been on the rewards that researchers can obtain by making their data open and usable. Researchers are used to getting credit for publishing papers about their research in academic journals, hence this mechanism is used to provide credit for publishing data. The mechanisms for data citation and publication are still under development, but early indications are that they will act as an incentive and encourage openness of data. For example, a survey of atmospheric science researchers carried out at the UK's National Centre for Atmospheric Science Conference in Bristol on 8–10 December 2008 showed that 67% of the 85 respondents agreed that they are more likely to deposit their data in a data centre if they can obtain academic credit through a data journal (Callaghan et al. 2009).

Publishing a Dataset in an Academic Context

Going back to the case study above, the GBS dataset differs from the ITALSAT dataset (and many others) in that it has been formally published in a data journal (Callaghan et al. 2013).

A data journal is an online journal that specialises in the publication of scientific data in a way that includes scientific peer-review. Most data journals publish short data papers cross-linked to, and citing, datasets that have been deposited in approved data centres.

A data paper is a short article that describes a dataset, and provides details of the dataset's collection, processing, software, file formats etc., allowing the reader to understand the when, how and why data was collected and what the data product is. The data paper does not require novel analyses or ground-breaking conclusions, instead the dataset is presented 'as is', allowing the publication of negative results.

Data journals support the development and enhancement of the scholarly record by providing a mechanism for:

- peer-reviewing datasets;
- publishing datasets quickly, as the data journal does not require analysis or novelty in the publication;
- providing attribution and credit for the data collectors who might not be involved with the analysis, and therefore would not be eligible for author credit for an analysis paper; and
- enabling the discovery and understanding of datasets, and providing assurance of their quality and provenance.

Data journals are becoming more prevalent in the scientific publishing ecosystem, signifying a recognition by publishers and funders that a mechanism for publishing data is required (and encouraging openness and access to data). For many researchers, who may be concerned that 'making their data open' is synonymous with 'giving it away and getting no credit', re-framing data sharing in the context of data citation and publication reassures them, and provides a structure and a framework that is well understood, where precedence and attribution are an accepted part of the publication and citation process.

There are many issues that need to be dealt with to ensure the smooth running of data journals, including (but not limited to) providing guidance to reviewers on how exactly to go about peer-reviewing a dataset, and how to certify that a data repository is suitably trustworthy for hosting published data. Data journals also rely significantly on a linking mechanism that is robust and reliable to link the article to the dataset, especially in those cases where the dataset is archived in a repository outside of the journal publisher's control. Linking between digital objects is commonplace on the Internet, but for the scholarly record to be maintained, the links between articles and datasets must be held to a higher standard of stability and reliability. These issues are not solved as of the date of this chapter, though there is a sizeable (and growing) community of researchers, librarians, data centre managers, academic publishers and research funders who are coming together to propose solutions and guidance for these problems.

Conclusions

Changing scientific culture is difficult and requires both incentives and disincentives, along with systems put in place to ease the process of change, and a critical mass of researchers who wish to make the change. The Earth and climate sciences have experienced their share of issues with lack of openness in the past (on a national level with Climategate, and on a multitude of personal levels, one example of which as described in this chapter). However, the push on researchers is definitely towards openness, and research funders are putting policies in place to support this. Bringing data into the academic publication process is potentially a very valuable way to encourage researchers to be more open with their data, while providing them with the credit they deserve for doing so.

References

British Atmospheric Data Centre 2007 National Database Of Atmospheric and Weather Data Tops 10,000 Users [British Academic Data Centre news release], 3 September. Available at https://badc.nerc.ac.uk/community/news/070906_Press.html [Last accessed 11 August 2014].

Callaghan, S A, Boyes, B, Couchman, A, Waight, J, Walden, C J and Ventouras, S 2008 An investigation of site diversity and comparison with itu-r recommendations. *Radio Science*, 43: RS4010. DOI:10.1029/2007RS003793.

Callaghan, S, Hewer, F, Pepler, S, Hardaker, P and Gadian, A 2009 Overlay journals and data publishing in the meteorological sciences. *Ariadne*, 60. Available at http://www.ariadne.ac.uk/issue60/callaghan-et-al/ [Last accessed 11 August 2014].

Callaghan, S A, Waight, J, Agnew, J L, Walden, C J, Wrench, C L and Ventouras, S 2013 The GBS dataset: measurements of satellite site diversity at 20.7 GHz in the UK. *Geoscience Data Journal*, 1(1): 2–6. DOI: 10.1002/gdj3.2.

Chappell, P R and Lorrey, A M 2013 Identifying New Zealand, Southeast Australia, and Southwest Pacific Historical weather data sources using Ian Nicholson's log of logs. *Geoscience Data Journal*, 1(1): 49-60. DOI: 10.1002/gdj3.1.

European Commission 2013 *Guidelines on Open Access to Scientific Publications and Research Data in Horizon 2020, Version 1.0*, 11 December. Available at http://ec.europa.eu/research/participants/data/ref/h2020/grants_manual/hi/oa_pilot/h2020-hi-oa-pilot-guide_en.pdf [Last accessed 11 August 2014].

Garcia-Herrera, R, Können, G P, Wheeler, D A, Prieto, M R, Jones, P D and Koek, F B 2005 CLIWOC: a climatological database for the world's oceans 1750–1854. *Climatic Change*, 73(1-2): 1-12. DOI: 10.1007/s10584-005-6952-6.

Guilyardi, E, Balaji, V, Lawrence, B, Callaghan, S, Deluca, C, Denvil, S, Lutenschlager, M, Morgan, M, Murphy, S and Taylor, K E 2013 Documenting Climate models and their simulations. *Bulletin of the American Meteorological Society*, 94(5): 623–627. DOI: 10.1175/BAMS-D-11-00035.1.

House of Commons Science and Technology Committee 2010 *Science and Technology Committee Eighth Report: The Disclosure of Climate Data from the Climatic Research Unit at the University of East Anglia*. Available at http://www.publications.parliament.uk/pa/cm200910/cmselect/cmsctech/387/38702.htm [Last accessed 11 August 2014].

National Science Foundation 2010 *NSF Data Sharing Policy.* Available at http://www.nsf.gov/bfa/dias/policy/dmp.jsp [Last accessed 11 August 2014].

Oliver, J and Kington, J A 1970 The usefulness of ships' log-books in the synoptic analysis of past climates. *Weather*, 25: 520–528. DOI: 10.1002/j.1477-8696.1970.tb04103.x.

Otung, I E, and Savvaris, A 2003 Observed frequency scaling of amplitude scintillation at 20, 40, and 50 GHz. *IEEE Transactions on Antennas and Propagation*, 51(12): 3259–3267. DOI: 10.1109/TAP.2003.820960.

Research Information Network 2008 *To Share or Not To Share: Publication and Quality Assurance of Research Data Outputs, Main Report.* Available at http://www.rin.ac.uk/system/files/attachments/To-share-data-outputs-report.pdf [Last accessed 11 August 2014].

The Royal Society 2012 *Science as an Open Enterprise. The Royal Society Science Policy Centre Report.* Available at http://royalsociety.org/uploadedFiles/Royal_Society_Content/policy/projects/sape/2012-06-20-SAOE.pdf [Last accessed 11 August 2014].

Research Councils UK 2013a *RCUK Policy on Open Access and Supporting Guidance.* Available at http://www.rcuk.ac.uk/documents/documents/RCUKOpenAccessPolicy.pdf [Last accessed 11 August 2014].

Research Councils UK 2013b *RCUK Policy on Open Access: Frequently Asked Questions.* Available at http://www.rcuk.ac.uk/RCUK-prod/assets/documents/documents/OpenaccessFAQs.pdf [Last accessed 11 August 2014].

Science and Technology Facilities Council, Chilbolton Facility for Atmospheric and Radio Research, [Callaghan, S A, Waight, J, Walden, C J, Agnew J and Ventouras, S] 2009a *GBS 20.7GHz Slant Path Radio Propagation Measurements, Sparsholt Site.* Available at http://badc.nerc.ac.uk/view/badc.nerc.ac.uk__ATOM__dep_11902946270621452 [Last accessed 11 August 2014]. DOI: 10.5285/E8F43A51-0198-4323-A926-FE69225D57DD.

Science and Technology Facilities Council, Chilbolton Facility for Atmospheric and Radio Research, [Callaghan, S A, Waight, J, Walden, C J, Agnew J and Ventouras, S] 2009b *GBS 20.7GHz Slant Path Radio Propagation Measurements, Chilbolton Site*. Available at http://badc.nerc.ac.uk/view/badc.nerc.ac.uk__ATOM__dep_11902119479621181 [Last accessed 11 August 2014]. DOI: 10.5285/639A3714-BC74-46A6-9026-64931F355E07.

Science and Technology Facilities Council, Chilbolton Facility for Atmospheric and Radio Research, [Callaghan, S A, Waight, J, Walden, C J, Agnew J and Ventouras, S] 2009c *GBS 20.7GHz Slant Path Radio Propagation Measurements, Dundee Site*. Available at http://badc.nerc.ac.uk/view/badc.nerc.ac.uk__ATOM__ACTIVITY_dc47dc7c-8880-11e1-9490-00163e251233 [Last accessed 11 August 2014]. DOI: 10.5285/db8d8981-1a51-4d6e-81c0-cced9b921390.

Science and Technology Facilities Council, Chilbolton Facility for Atmospheric and Radio Research, [Ventouras, S, Callaghan, S A and Wrench, C L] 2012a *ITALSAT Radio Propagation Measurement at 20GHz in the United Kingdom*. Available at http://badc.nerc.ac.uk/view/badc.nerc.ac.uk__ATOM__ACTIVITY_e6d8b012-a65d-11e1-94b7-00163e251233 [Last accessed 11 August 2014]. DOI: 10.5285/3158D138-D592-4045-ADE4-B76CF9F42129.

Science and Technology Facilities Council, Chilbolton Facility for Atmospheric and Radio Research, [Ventouras, S, Callaghan, S A and Wrench, C L] 2012b *ITALSAT Radio Propagation Measurement at 40GHz in the United Kingdom*. Available at http://badc.nerc.ac.uk/view/badc.nerc.ac.uk__ATOM__ACTIVITY_52ad3c54-a663-11e1-ba03-00163e251233 [Last accessed 11 August 2014]. DOI: 10.5285/4A60EE2F-0FD1-4141-9244-7BEBF240BB49.

Science and Technology Facilities Council, Chilbolton Facility for Atmospheric and Radio Research, [Ventouras, S, Callaghan, S A and Wrench, C L] 2012c *ITALSAT Radio Propagation Measurement at 50GHz in the United Kingdom*. Available at http://badc.nerc.ac.uk/view/badc.nerc.

ac.uk__ATOM__ACTIVITY_f2984bd6-a664-11e1-ac44-00163e251233 [Last accessed 11 August 2014]. DOI: 10.5285/597C906A-B09E-4822-8B60-3B53EA8FC57F.

US Geological Survey 2000 *USGS Fact Sheet-036-00 March 2000*. Available at http://pubs.usgs.gov/fs/2000/fs036-00/fs036-00.pdf.

Ventouras, S, Callaghan, S A and Wrench, C L 2006 Long-term statistics of tropospheric attenuation from the Ka/U band ITALSAT satellite experiment in the United Kingdom. *Radio Science*, 41: RS2007. DOI:10.1029/2005RS003252.

Open Minded Psychology

Wouter van den Bos, Mirjam A. Jenny
and Dirk U. Wulff

Center for Adaptive Rationality, Max Planck Institute
for Human Development, Berlin, Germany

Introduction

Psychology is a young and dynamic scientific discipline, which has a history of closely scrutinizing its own methods. For example, in the sixties, experimental psychology improved its methods after researchers became aware of the experimenter effect, that is, experimenters may inadvertently influence experimental outcomes (Kintz et al. 1965). The introduction of new technologies such as neuroimaging in the late nineties also raised several unique methodological issues (e.g. reverse inferences and double dipping: Poldrack, 2006; Kriegeskorte et al., 2009). Finally, debating and improving our statistical toolbox has always been

How to cite this book chapter:
van den Bos, W., Jenny, M. A. and Wulff, D.U. 2014. Open Minded Psychology. In: Moore, S. A. (ed.) *Issues in Open Research Data*. Pp. 107–127. London: Ubiquity Press. DOI: http://dx.doi.org/10.5334/ban.g

an integral part of the field: many psychology departments have methods departments and there are several dedicated journals (e.g. *Behavior Research Methods* since 1969). Currently, advancements of online technologies hold the potential to transform the field regarding the reporting and sharing of data.

There has been a shift from "paper only" to the digital presence of scientific journals, which has lifted the physical limits of research reports, allowing for publication of much more extensive supplementary material. Open Access journals like those of PLOS and *Frontiers* are on the rise, and open access options become the standard. Finally, online repositories and collaborative tools (e.g. openscienceframework.org) allow for effortless and free storing of data, experimental designs, analysis code, and additional information needed for successful replication or meta-analyses.

Psychology as a field has always been quick to integrate new technologies into their experimental design and measurement, such as computerized experiments and neuroimaging techniques. However, as David Johnson observed in 2001, "psychological science has largely taken a pass on optimizing knowledge production and integration through use of electronic communication" (Johnson 2001). Now almost 15 years later, with a few notable exceptions, this still largely rings true. To openly share material, over the past decades physicists, mathematicians, and computer scientists used arXiv.org; molecular biologists used the Protein Data Bank; and GenBank, geoscientists, and environmental researchers have Germany's Publishing Network for Earth and Environmental Science (PANGAEA). However, nothing of the like has been developed in psychology.

As a result, the collection of data is still surprisingly cumbersome. According to one study, around 73% of corresponding authors failed to share data from their published papers upon request (Wicherts 2013) Luckily, one of our authors has been more successful;

For our meta-analyses, we aimed to combine the results of 20 papers, all published between 2004 and 2013 (Wulff, Hertwig & Mergenthaler, in prep.). Out of those papers, 16 were first-authored by 11 different researchers outside our own research group and thus needed formal contacting to request the data. After first contact in April 2012, it took nearly three months (84 days) to either completely retrieve the requested data or have certainty over its unavailability (three datasets). A total of 68 emails were exchanged and 12 reminders needed to be sent (see **Figure 2**).

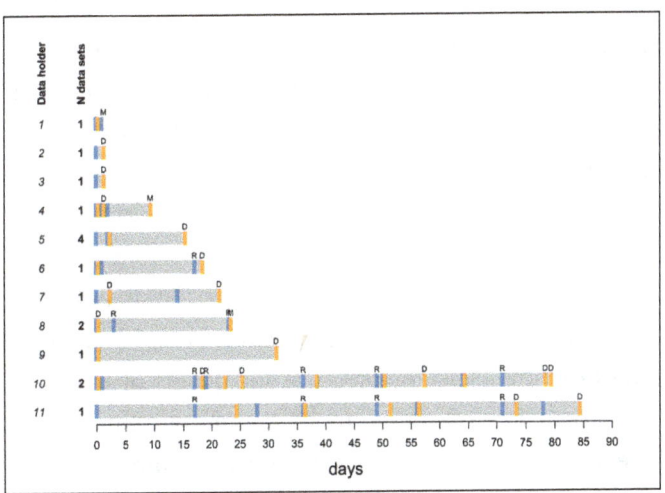

Figure 2: Correspondence timeline for retrieving a total of 16 data sets from 11 data holders.

D = Data; M = Missing (data that was eventually declared missing); R = Reminder. Blue marking indicates emails of the requester, orange of the data holder. Multiple data indicators may result from requests for multiple datasets, but also from incomplete data submission.

(Box continued on next page)

(Box continued from previous page)

> The two main reasons for the delay seemed to be the unavailability of the data for the researcher herself and low prioritization. An exemplary response was this: "I'm quite busy for the next few days but will send this on to after that. Please remind me again by the end of the month." Like this one, most correspondences made very apparent that providing data meant a substantial amount of work for the providing researcher. This was further illustrated by three cases where data was provided in separate chunks or required later supplement due to initially incomplete data.
>
> A second and often more bothersome obstacle arose after the data was retrieved: bringing the data into a coherent organization scheme. For (now) obvious reasons, this work remains with the requester. Usually data has not only been collected in slightly different paradigms and with different tools, they also come in different formats (e.g. long or wide). The restructuring requires a lot of manual labor, but also a significant amount of intellectual work to understand the data structure. Here, the presence and quality of accompanying data documentation took an important moderating role. In the study, the level of documentation ranged from not being there to elaborate and easily intelligible descriptions of how the data correspond to the elements of the published paper. Clearly, some of the descriptions were crafted for this instance, which pointed again to the merits of making documented data available upon publication.

Box 1: Meta-analyses: A Case Study.

he managed to receive over 85% of the requested datasets for this meta-analysis. However, collection of only 16 datasets took about three months, and this does not include the time spent on the subsequent organization of the data for analysis (see **Box 1**). Had these datasets been stored in a repository in a standardized format, their collection would probably have taken five minutes (which is approximately 25,000 times faster). Such slow and incomplete dataset collection clearly hinders academic progress. For this and other reasons, there is a growing call for increased openness in sharing data in psychology (Miguel et al. 2014; Pitt & Yang 2013; Wicherts 2013).

In this chapter we aim to make a case for the need of a common data sharing policy for psychological science, discuss what such a policy should address, and hope to make some practical suggestions along the way. First, we summarize the reasons for open data and what the advantages could be specifically for psychological science. Next, we will address in more detail what it means for data to be truly open, as well as some concerns about open data. Finally, we discuss how we could move toward a more open minded psychology.

Why Open Data?

One argument for open data that has received a lot of attention recently has been a number of cases of data fraud in science. Although it is likely that open data requirements may reduce fraudulent behavior (Simonsohn 2013), we do not think that an open data policy should be based on the motivation of exposing fraudulent behavior. Instead we deem it more successful to highlight the numerous benefits of data sharing, in general and for psychology specifically.

To start with a very straightforward benefit, data sharing leads to better data preservation. Technological advancements (or planned obsolescence) quickly make our storage media obsolete and unusable (floppy drive anyone?), rendering the data stored on it inaccessible. In addition, scientists move locations in many stages of their careers, each time introducing the danger of the data getting lost. Of course, many researchers think this would not happen to them, but the results of published data requests do suggest that lost data is probably one of the main causes of non-compliance. Luckily, most online repositories have structured institutional funding and make use of professional servers that provide continuous backups of stored data. As such, there really is no reason for data to get lost; it can now be potentially stored forever.[1]

Crucially, when data is openly available it can be used in many ways; it can be combined with other datasets, used to address questions that were not thought of by the authors of the original studies, analyzed with novel statistical methods that were not available at the time of publication, or used as an independent replication dataset.

One very successful example of such a project is the 1000 Functional Connectomes Project.[2] This project, explicitly modeled upon the successful collaborative efforts to discover the human genome, was formed to aggregate existing resting state functional magnetic resonance imaging (R-fMRI) data from collaborating centers throughout the world.[3] The initiators of this project were

---

[1] This is a lot longer than the mere five years that is currently indicated in the publication manual of the American Psychological Association (APA manual sixth edition) as a reasonable time to keep your data, more on this below.

[2] See: http://fcon_1000.projects.nitrc.org/

[3] Imaging the brain during rest reveals large-amplitude spontaneous low-frequency (<0.1 Hz) fluctuations in the fMRI signal that are temporally correlated across functionally related areas. Referred to as functional connectivity, these correlations yield detailed maps of complex neural systems, collectively constituting an individual's "functional connectome."

able to gather and combine R-fMRI data from over 1200 volunteers collected independently at 35 centers around the world (Biswal et al. 2010). Using this large dataset, the researchers were able to establish the presence of a universal functional architecture in the brain and explore the potential impact of a range demographic variables (e.g. age, sex) on intrinsic connectivity. An additional benefit from such a collaborative effort is the size of the dataset that is created in the process. Due to high costs and limited access to facilities, studies in the cognitive neurosciences currently have rather small sample sizes (Button et al. 2013), which may result in overestimates of effect sizes and low reproducibility of results. Thus, combining efforts to create larger sample sizes would be one way to address this issue.[4] Of course, re-analysis may also entail much more straightforward secondary analyses such as those that may be raised during the review process (e.g. how about using variable X as a covariate?), which may provide readers with more insight into the impact of published results (Pitt & Tang 2013).

Finally, science is and should be a cumulative enterprise. Optimally, scientists digest the cumulated relevant literature and incorporate the extracted knowledge in designing their own new and potentially better experiment. Still, a single experimental setup is often repeated by different scientists under only mildly different conditions. In supplement to the accumulation of theoretical knowledge, such repetitions enable an interested researcher to actively cumulate existing evidence by means of combined statistical analyses, i.e. meta-analyses. Although meta-analyses can be

---

[4] It is commonly believed that one way to increase replicability is to present multiple studies. If an effect can be shown in different studies, even though each one may be underpowered, many will conclude that the effect is robust and replicable. However, Schimmack (2012) has recently shown that this reasoning is flawed.

done on group level statistics that are extracted from papers, such as effect sizes or foci of brain activity, there are several benefits to using the raw data for meta-analyses (Cooper & Pattall 2009; Salimi-Khorshidi et al. 2009). As we pointed out in our example (**Box 1**), open data would greatly facilitate meta-analyses.

Of course, data could also be used for purposes other than analysis. For instance, data can be used in courses on statistics and research methods (Whitlock 2011), as well as in the development and validation of new statistical methods (Pitt & Tang 2013).

Finally, one could argue that the results of publicly funded research should, by definition, be made publicly available. Reasoning along these lines, many funding bodies are increasing the degree to which they encourage open archiving. However, there should not be two classes of data: publicly funded open access and privately funded "hidden" datasets. Ideally, all data should be publicly available at the latest after the first study with them has been published.

What Is Open Data?

> A piece of data or content is open if anyone is free to use, reuse, and redistribute it—subject only, at most, to the requirement to attribute and/or share-alike.
> (The Open Knowledge Foundation)

This is the short version of the Open Definition provided by the Open Knowledge Foundation.[5] Although open data sounds very straightforward, it may actually be more complicated than you think. Open data does not just mean storing your data on your personal website. For it to be open, the data also need to be usable by others. There are several criteria that should be met for data to be truly usable.

---

[5] ofkn.org; for the full length definition see: http://opendefinition.org/

First and foremost, other people need to be able to find the data; it should be discoverable. Archiving data at online repositories (see **Box 2**) significantly increases discoverability of data. These repositories will most likely be around longer than personal websites and often also allow for the storage of additional materials (e.g. code). Many of these repositories have good search functions so related datasets will be found with a single search. As an added bonus, several online repositories, like figshare, provide all datasets with a DataCite digital object identifier (DOI). As a result, these datasets can be cited using traditional citation methods and citations can be tracked. In fact, Thomson Reuters has recently launched a Data Citation Index.[6]

Second, if your data is discoverable, it should also be usable. As pointed out in our own meta-analyses case study, usability can be a matter of degree. Psychologists make use of a whole suite of different software tools to collect their data, many of which are proprietary such as MATLAB or E-Prime, or are dependent on proprietary software such as SPM or psychtoolbox. Subsequently, data is often also organized in proprietary software packages such as SPSS, SAS or Microsoft Excel. Output files from these software packages are not truly open because you first need to buy a (sometimes) expensive program to be able to read them. Currently, not many psychologists seem to be aware of this. To illustrate this, we made a random draw of two 2013 issues of the journal of the Society for Judgment and Decision Making, a journal with probably the best data sharing culture in psychology. It revealed that more than two thirds of the shared data was in proprietary format.[7] The solution here is simple, all of the software packages have the

---

[6] See: http://thomsonreuters.com/data-citation-index/
[7] Datasets of issues 1 and 2 of 2013 in order of frequency: 6 SPSS, 5 CSV, 3 Excel, 1 STATA, and 1 MATLAB.

Where to share your data

Repositories

- openfmri.org
- figshare.com
- openscienceframework.org
- psychfiledrawer.org

Data publications

- PsychFileDrawer
- *Journal of Open Psychology Data*
- Nature's *Scientific Data*

Licensing your data

When licensing your data, *JOPD* recommends any of these for licenses:

- Creative Commons Zero (CC0)
- Open Data Commons Public Domain Dedication and License (PDDL)
- Creative Commons Attribution (CC-By)
- ODC Attribution (ODC-By)

All of the above licenses carry an obligation for anyone using the data to properly attribute it. The main differences are whether this is a social requirement (CC0 and PDDL) or a legal one (CC-By and ODC-By). The less restrictive

(Box continued on next page)

(Box continued from previous page)

> your license, the greater the potential for reuse. In general, it is not recommended to use licenses that impose commercial or other restrictions on the use of data (for more on licensing see Chapter 3.

Box 2: Putting Your Data Online—a Practical Guide.

option to export the data to formats that are readable by every machine or operating system (e.g. CSV or TXT).

Next, for data to be usable, it must be completely clear how to read the data files. When a published paper is accompanied by open access data it may be easy to understand the content of the data file just from the header information. However, for some more complex datasets, such as neuroimaging data, this may not be the case. In this instance, it is important to make sure that others can use the data. Of course good standards for structuring complex datasets further increases usability (e.g. OpenfMRI standards for fMRI data https://openfmri.org/).

Finally, it is important to license your data when you share it to make sure it is as open as you want. When there is no license, it is not clear to what extent the data is open, and it is thus effectively unusable. Luckily, Creative Commons and the Open Knowledge Foundation made it very easy for scientists to decide how open they want their data to be. In addition, the *Journal for Open Psychology Data (JOPD)* set up recommendations for psychologists (see **Box 2** for more information). To summarize, these licenses make sure that people can use the data but also obligates users to properly attribute it.

Concerns: Privacy and consent

When dealing with personal data on public repositories it is of the utmost importance to protect the privacy of the participants. Of course there are already very good rules in place, but a mistake is easily made and with the development of new technologies (and, ironically, more open data) it becomes easier than ever to identify persons from just small pieces of data. For behavioral experiments in psychology it often seems enough just to replace participants' names with codes. But here is one sobering statistic from the Sweeney's Data Privacy Lab: About half of the US population (53%) are likely to be uniquely identified by only place, gender, or date of birth. This goes up to 87% when place is specified as a zip code (Sweeney 2000).[8] Date of birth and gender are of course very general measures, and place can often be derived from the university where the researchers are based. An interest in social economic status may lead researchers to store zip codes too. It is customary to report age at time of the study instead of birth date, but this illustrates the consequences when dates of birth are accidentally shared.

The increasing use of biological measures (brain, hormones, DNA) in psychology not only further increases the challenge to keep participants data anonymous but also makes anonymous data storage more pressing. For instance, when submitting brain imaging data to a public repository it is very important to extensively de-identify your images. There are three important sources of identifiable information. The most obvious is of course the filename. Less obvious are the information stored in the file headers, and the three-dimensional image of the participant's face that is

---

[8] If you live in the US we encourage you to check this out for yourself using the Sweeney's web app at http://aboutmyinfo.org/

often part of the image (see **Figure 1**). Luckily, there are several tools that help remove this data (e.g. LONI De-identification Debablet). However, de-anonymization for other types of biological data can be more difficult, maybe even impossible. Single gene

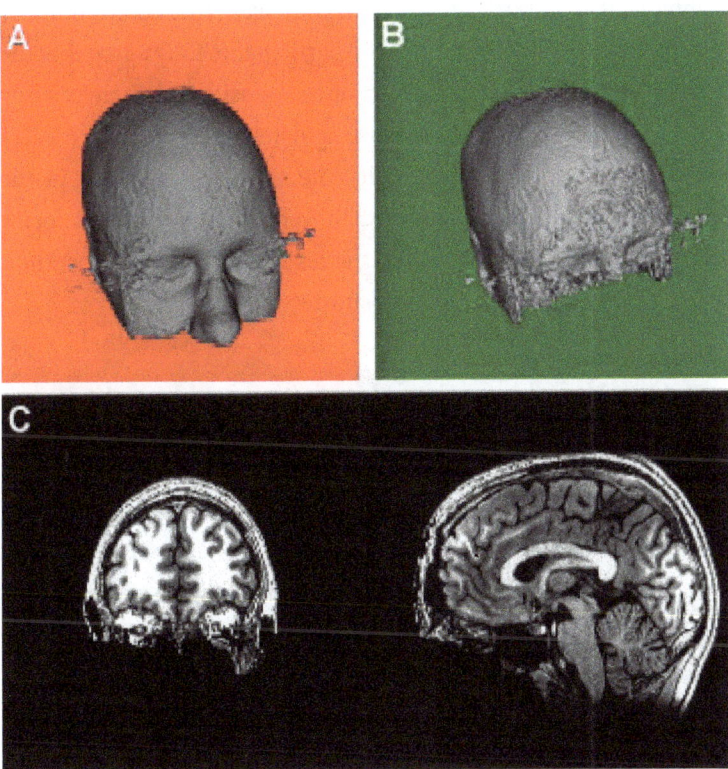

Figure 1: Structural Image before (A) and after defacing (B). It is clear that facial features are no longer recognizable but (C) essential brain data is still fully accessible. Removing the face of a structural image can be done using the Mbrin defacer package (http://www.nitrc.org/projects/mri_deface/). Alternatively, FSL brain extraction tool could be used to remove all non-brain tissue, although this might be more time intensive. In both cases is it essential to check whether deidentification is successful while brain data is still 100% intact.

mutation data is not very revealing, but genome-wide association data, or even just parts of it, can lead to identifiable data. At the moment it is not even clear whether such data can ever be shared anonymously (Hayden 2013).

Finally, it is important that both the local ethics committee or institutional review board and the participants fully agree that the data is eligible for posting on an open repository. Whereas some ethical committees allow for existing data to be posted in repositories, most will require explicit statements to this effect in the consent form. Thus, to enable data to be shared, consent forms must inform participants about the researcher's wish to post their anonymous data on public repositories.[9]

Quo Vadis? Intentions, Integration and Incentives

One of the authors recently showed that psychologists appear to be less in favor of mandatory conditions of publication than standards of good practice (Fuchs, Jenny & Fiedler 2012). Before we answer the question if good standards will suffice, let us first examine the existing standards for psychology.

In most fields, the standards are set by the journals, grant agencies and professional societies or associations. In their publication manual, the American Psychological Association (APA) encourages the open sharing of data (APA 6th ed., p.12), but has surprisingly little to say on the matter. The whole section is not even a page and no longer seems to be in line with current sentiments in the field. It mainly suggests several limiting factors on sharing, and relieves the publishing scientist of most responsibilities. First,

---

[9] Because of the potential risk with DNA data, these consent forms have become very long and include mandatory examinations to make sure all details are understood.

there is the surprisingly short period of five years that researchers are recommended to keep their data. Next, APA suggests that sharing is done only with qualified researchers (?), that all the costs should be borne by the requester, and they stress that both parties sign a written agreement that further limits data use. There is no mention of public repositories, data formatting or licensing.

In general, most journals have followed guidelines similar to those suggested by the APA. That is, they encourage but do not require data sharing.[10] However, as we mentioned, this encouragement has not resulted in data sharing on public repositories, and direct data requests are not met with great enthusiasm. Thus, on several issues, good standards are absent, and current guidance has not spurred researchers to publish or even share their data. In other words, there is room for a new data sharing policy for psychological science, one that is even more open minded. What should such a policy look like? Such a policy should of course represent the whole community. And it should also seriously address the concerns of its constituents, the scientists. Here, we briefly address issues that we think should be considered.

Concerns

One of the main concerns is that other researchers might publish research done with the data that the original researchers were also planning to work on. This worry applies to longitudinal studies especially. We view this as a valid and understandable concern and would like to propose a few conditions for the publication of data, which could resolve these worries. As long as research is still

---

[10] Exceptions being the *Journal of Cognitive Neuroscience* (2000 to 2006) and *Judgment and Decision Making*.

ongoing, authors could define embargoes or publish only part of the data. How such embargoes would be set up would have to be agreed upon in the community, however, and the embargoes should not hinder research progress. As is currently the case with patents, there could be a fixed term after which data could no longer be embargoed. Building on the existing guidelines, the rule could be that after the five-year period suggested by APA for researchers to keep their data, the data must then be shared. This embargoed period preferably should be shorter. Another concern is copyright for data that was funded privately. Here, there is probably no single solution that would always be effective, but research institutions should negotiate to make the data publicly available whenever possible.

More importantly, research in other fields (Pitt & Tang 2013) suggests that one major reason for scientists not being willing to share their data is because it is too much work. But once data sharing is the norm and researchers plan ahead, this argument does not hold. Yes, searching for old data on a bunch of hard disks, CD-ROMs or outdated laptops may be tedious, but usually researchers have easy access to their own data. Organizing the data in a self-explanatory fashion poses a bit of extra work but also further ensures data quality as the data is double-checked. Furthermore, platforms such as the Open Science Framework are extremely helpful for organizing, storing and making your data accessible. Together with a good data management plan, which should be taught at the latest in graduate school, data sharing can be made quick and simple.

Incentives—badges, data publication, citations—versus enforcement

Even if sharing data is not a large time investment for most (early career) scientists, the time can be better spent writing a paper.

So it seems that even if many scientists agree that data should be made openly available, and sharing data can be done almost automatically, the right incentives are still required to get them to actually do it.

First of all there is of course enforcement. Journals, especially high impact journals, and grant agencies could simply make open data obligatory. Following other fields, psychology could simply force open data on itself. Although enforcement is probably the most effective it is the least attractive strategy (Fuchs, Jenny & Fiedler 2012) and we therefore would like to consider some alternatives.

Recently, several journals (including flagship *Psychological Science*) adopted the badges provided by the Center for Open Science to further encourage data sharing. If authors post their data and other material online, they receive an open data badge. However, it is unclear how much improvement these badges will bring given that researchers are mostly evaluated by high impact papers and the number of citations. Of course, the journals could make the badges more powerful by, for example, ensuring increased attention or promotion for badged articles. In addition, grant agencies could amplify the effect of badges by taking them into consideration when evaluating grant applications. Another more traditional way of encouraging open data that is currently being implemented is turning data sharing into citable data publications (**Box 2**). Finally, it is worth pointing out that several studies have now shown there is a general citations advantage for papers that are accompanied by open data (Piwowar & Vision 2013).

Data storage

It must be clear who is responsible for the storage of and access to the data. The publications must indicate where

> the raw data is located and how it has been made permanently accessible. It must always remain possible for the conclusions to be traced back to the original data.
> (Levelt, Noort and Drenth Committees, 2012, p.58)

This quote from the report on the fraudulent psychologist Diederik Stapel highlights two important issues that need to be addressed. First, it suggests that data should be permanently accessible, which is a much more than the five years the APA currently recommends. We do agree that we should opt for much longer data retention; however, this raises the question of how long exactly (given permanently means forever, and that is a mighty long time). Second, it raises the question of who is responsible for storage and what are sustainable models for storing data for such long periods of time? Currently some of the online repositories are commercially funded whereas others are backed by universities. For instance, figshare is backed by the company Digital Science, but what happens with the data if Digital Science goes bankrupt? How should the costs of sharing data (in terms of time and money) be distributed?

Conclusion

For truly open science, not only should data be openly accessible but also the code for experiments and data analysis. This makes it easier to completely understand and replicate analyses, and prevents researchers from having to repeat the same workload by reprogramming an already existing task. As with the data, the codes should ideally be published in openly available languages such as R and Python (e.g. using PsychoPy or OpenSesame).

To make psychological data available and openly accessible, we must work toward a shift in researchers' minds. Open science is simply more efficient science; it will speed up discovery and our understanding of the world. It is good to remind ourselves of this bigger picture when we are writing papers and grant proposals. We think the time is ripe for more open minded psychology, and hope with this chapter we contribute to the ongoing discussion and work toward a common data policy, and at the same time we have tried to point out several tools that are already available.

References

American Psychological Association 2010 *Publication Manual of the American Psychological Association* (6th ed.). Washington DC: APA.

Button, K S, Ioannidis, J P, Mokrysz, C. Nosek, B A, Flint, J, Robinson, E S and Munafò, M R 2013 Power failure: why small sample size undermines the reliability of neuroscience. *Nature Reviews Neuroscience,* 14(5): 365–376. DOI: 10.1038/nrn3475.

Biswal, B B, Mennes, M, Zuo, X N, Gohel, S, Kelly, C, Smith, S M, Beckmann, C F, Adelstein, J S, Buckner, R L, Colcombe, S, Dogonowski, A M, Ernst, M, Fair, D, Hampson, M, Hoptman, M J, Hyde, J S, Kiviniemi, V J, Kötter, R, Li, S J, Lin, C P, Lowe, M J, Mackay, C, Madden, D J, Madsen, K H, Margulies, D S, Mayberg, H S, McMahon, K, Monk, C S, Mostofsky, S H, Nagel, B J, et al. 2010 Toward discovery science of human brain function. *Proceedings of the National Academy of Sciences of the United States of America*, 107(10): 4734–4739. DOI:10.1073/pnas.0911855107.

Cooper, H and Patall, E A 2009 The relative benefits of meta-analysis conducted with individual participant data versus aggregated data. *Psychol Methods,* 14(2): 165–176. DOI: 10.1037/a0015565.

Fuchs, H M, Jenny, M and Fiedler, S 2012 Psychologists are open to change, yet wary of rules. *Perspectives on Psychological Science,* 7(6): 639–642. DOI:10.1177/1745691612459521.

Hayden, E C 2013 Privacy protections: the genome hacker. *Nature,* 497(7448): 172–174. DOI: 10.1038/497172a.

Johnson, D H 2001 Sharing data: it's time to end psychology's guild approach. *APS Observer,* 14(8). Available at https://www.psychologicalscience.org/index.php/uncategorized/sharing-data-its-time-to-end-psychologys-guild-approach.html [Last accessed 12 August 2014].

Kintz, B L, Delprato, D J, Mettee, D R, Persons, C E and Schappe, R H 1965 The experimenter effect. *Psychological Bulletin,* 63(4): 223–232.

Kriegeskorte, N., Simmons, W. K., Bellgowan, P. S. F., & Baker, C. I. (2009). Circular analysis in systems neuroscience: the dangers of double dipping. *Nature Neuroscience, 12*(5), 535–540. Retrieved from http://www.nature.com/doifinder/10.1038/nn.2303.

Levelt, Noort and Drenth Committees 2013 *Flawed Science: The Fraudulent Research Practices of Social Psychologist Diederik Stapel.* Available at https://www.commissielevelt.nl/wp-content/uploads_per_blog/commissielevelt/2013/01/finalreportLevelt1.pdf [Last accessed 12 August 2014].

Miguel, E, Camerer, C, Casey, K, Cohen, J, Esterling, K M, Gerber, A, Glennerster, R, Green, D P, Humphreys, M, Imbens, G, Laitin, D, Madon, T, Nelson, L, Nosek, B A, Petersen, M, Sedlmayr, R, Simmons, J P, Simonsohn, U and van der Laan, M 2014 Promoting transparency in social science research. *Science,* 343(6166): 30–31. DOI: 0.1126/science.1245317.

Pitt, M A and Tang, Y 2013 What should be the data sharing policy of cognitive science? *Topics in Cognitive Science,* 5(1): 214–221. DOI:10.1111/tops.12006.

Piwowar, H A and Vision, T J 2013 Data reuse and the open data citation advantage. *PeerJ,* 1: e175. DOI: 10.7717/peerj.175.

Poldrack, R. (2006). Can cognitive processes be inferred from neuroimaging data? *Trends in Cognitive Sciences, 10*(2), 59–63. Retrieved from http://linkinghub.elsevier.com/retrieve/pii/S1364661305003360.

Salimi-Khorshidi, G, Smith, S M, Keltner, J R, Wager, T D and Nichols, T E 2009 Meta-analysis of neuroimaging data:

a comparison of image-based and coordinate-based pooling of studies. *NeuroImage*, 45(3): 810–823. DOI:10.1016/j.neuroimage.2008.12.039.

Schimmack, U 2012 The ironic effect of significant results on the credibility of multiple-study articles. *Psychological Methods*, 17(4): 551–566. DOI:10.1037/a0029487.

Simonsohn, U 2013 Just post it: the lesson from two cases of fabricated data detected by statistics alone. *Psychological Science*, 24(10):1875–1888. DOI:10.1177/0956797613480366.

Sweeney, L 2000 *Simple Demographics Often Identify People Uniquely. Carnegie Mellon University, Data Privacy Working Paper 3*. Pittsburgh PA: Carnegie Mellon University. Available at http://impcenter.org/wp-content/uploads/2013/09/Simple-Demographics-Often-Identify-People-Uniquely.pdf [Last accessed 12 August 2014].

Wicherts, J M 2013 Science revolves around the data. *Journal of Open Psychology Data* 1(1): e1. DOI: 10.5334/jopd.e1.

Whitlock, M C 2011 Data archiving in ecology and evolution: best practices. *Trends in Ecology and Evolution*, 26(2): 61–65. DOI:10.1016/j.tree.2010.11.006.

Open Data in Health Care

Tom Pollard[*] and Leo Anthony Celi[†]

[*]University College London (UCL), London, UK
[†]Massachusetts Institute of Technology (MIT), Cambridge, USA

Signs of Life

As we pass through life in the digital era we leave a health trajectory in our wake. Phones, shopping habits, and visits to the doctor create a trace of data that can be used to not only assess our past and present wellbeing, but also forecast the future. To some, this is an unparalleled opportunity to improve health care, whereas to others it is an emerging threat to civil liberty. Most of us camp somewhere between the two poles: we see the rewards and we acknowledge the concerns. The question is how we move past this point, when business models and legal frameworks, built for

---

How to cite this book chapter:
Pollard, T. and Celi, L. A. 2014. Open Data in Health Care. In: Moore, S. A. (ed.) *Issues in Open Research Data*. Pp. 129–140. London: Ubiquity Press.
DOI: http://dx.doi.org/10.5334/ban.h

a pre-internet world, struggle to keep up with the pace of change (Park & VanRoekel 2013).

The movement to give us open access to research articles began roughly fifteen years ago[1]. Before the dust has settled, there is now a strong push from researchers, funders, and publishers to open the data that underpins those articles. The suggestion to share research data is hardly new—Sir Francis Galton entertained this thought in 1901 (Hrynaszkiewicz & Altman 2009)—but technology now exists to enable sharing with relative ease. Culture is largely the barrier that restricts flow of research data, and for data sharing to be adopted there are challenges to overcome around privacy, competition, and incentives to share (Wellcome Trust 2014).

Improving Care

The medical and biomedical research professions have come under heavy criticism in recent years (Celi 2014). The Institute of Medicine's 1999 report 'To err is human', for example, estimated that between 44,000 people and 98,000 people die in US hospitals each year as a result of preventable medical errors, with even the lower estimate exceeding mortality of threats such as AIDS and breast cancer (eds. Kohn, Corrigan & Donaldson 2000). Further high profile blows were delivered in 2013, with the US National Research Council's report on 'Shorter Lives, Poorer Health' and The Economist's 'Unreliable research: Trouble at the lab' (National Research Council 2013; The Economist, 2013). 'Half of what we know might be wrong, and the other half useless,' is perhaps the

---

[1] Two key reference points are Steven Harnad's 'subversive proposal' in 1994 and the founding of the Budapest Open Access Initiative in 2001.

most damning appraisal of the state of medical knowledge, coming from Professor John Ioannidis in his editorial 'How Many Contemporary Medical Practices are Worse than Doing Nothing or Doing Less?' (Ioannidis 2013).

It is widely acknowledged that better handling of information could address many of the criticisms, potentially helping to transform the quality of research and care (Institute of Medicine 2000; Wellcome Trust 2014). When data is not shared, quality of care suffers through inefficiencies, proliferation of errors, and wasted opportunities for learning. An open approach enables refinement of knowledge and collaborative growth towards united goals (Ioannidis et al. 2014).

When efforts are collaborative, progress can be rapid. One such example was the global research effort in 2011 to sequence and analyse the genome of a toxic strain of *Escherichia coli*, quickly helping to control the outbreak and prevent further deaths (The Royal Society 2012; Rohde et al. 2011). Transparency, through open data, can also highlight potential cost savings in our health systems. A recent study in England suggested potential savings of over £300 million pounds per year through switching to generic equivalents of two branded drugs (Allen 2012).

Qualifying 'Open'

The definition of open data is unequivocal: 'A piece of data is open if anyone is free to use, reuse, and redistribute it—subject only, at most, to the requirement to attribute and/or share-alike' (Open Knowledge, 2013). This is a copyright-centric model of sharing, facilitated by the adoption of 'copyleft' licences that allow reproduction and reuse (Hrynaszkiewicz & Cockerill 2012; Korn & Oppenheim 2011). This approach to sharing means there are few downstream restrictions, allowing, for example, reuse in classrooms, industry, research, and 'citizen science', maximising the potential of the data.

Where we are dealing with sensitive information, as we often are in health, it is fair to accept that there is a limit to what can be shared openly. Unless explicit consent for sharing has been obtained, details may have to be abstracted or removed to protect the individuals. Finding the appropriate balance between anonymisation and retaining useful detail is not straightforward, often involving a trade-off between risk and value.

As a result of this trade-off, John Wilbanks, who worked for years at Creative Commons[2], suggests that the copyright-centric approach may be unsuitable for health data. Wilbanks champions an alternative model built on trust (Howard 2012). Projects that have adopted this privacy-centric approach include his Portable Legal Consent study and Sage Bionetworks' clinical research studies, which seek to match participants willing to share their data with networks of researchers under contract to 'play fair' by returning research insights and not attempting to re-identify individuals.

It is likely and desirable for data sharing to progress on both privacy- and copyright-centric branches: we will get better at sharing 'true' open data with few restrictions on downstream reuse, and we will also develop platforms for sharing within trusted networks. Complementing both approaches are practical measures of openness, which assess whether data can be found, accessed, and reused. Open Knowledge has assembled a list of examples of 'bad data', which emphasise, lightheartedly, that there is more to sharing data than dropping files onto

---

[2] Creative Commons: http://creativecommons.org/

a public website (Open Knowledge, 2014a). Another initiative by Open Knowledge, the Open Data Index, provides a series of questions to assess the availability and openness of data, asking, for example, whether data is machine readable (e.g. text instead of image), available in a non-proprietary format (e.g. CSV instead of XLS for tabulated data), and openly licensed (e.g. with a Creative Commons licence) (Open Knowledge, 2014b).

Doing No Harm

Radicals may be prepared to bare all on the web, but the majority of us have expectations that certain information will remain within trusted networks. Our desire for privacy goes beyond avoidance of embarrassment. Revealing identifiable information that relates to our physical, mental, and social wellbeing has risks, for example by enabling discrimination by insurers or employers. While the level of risk can be debated and varies from case to case, it is clear that damage is possible. In a well-referenced case in 2008, for example, a nurse's career was compromised when confidential health information was leaked to her employer (European Court of Human Rights 2008).

All health data is sensitive and should be treated with respect, but the specific legal provisions that regulate data processing and sharing vary by location (UK Parliament 1998; United States Congress 1996). Regardless of the legal framework, regulation is implemented to achieve a similar effect—protection of the data subjects—and so rather than discuss detail in specific locations we give an introduction to the general concepts here. Our aim is to protect the individual, and so whether or not an item is defined as 'personal information' we should err on the side of caution when sharing data to mitigate the risks of harm. Open data must either not identify the individual or there must be explicit consent to share.

Anonymisation is a method that can be employed to open up health data, by separating information from the individual. The EU Data Protection Directive, for example, states that 'the principles of protection shall not apply to data rendered anonymous in such a way that the data subject is no longer identifiable. In addition, the UK Information Commissioner's Office's Anonymisation: Managing Data Protection Risk Code of Practice document notes:

> There is clear legal authority for the view that where an organisation converts personal data into an anonymised form and discloses it, this will not amount to a disclosure of personal data. This is the case even though the organisation disclosing the data still holds the other data that would allow re-identification to take place.
> (Information Commissioner's Office 2012)

Successful anonymisation is not straightforward, however, and there are examples of both failure and success (El Emam et al. 2012; Neamatullah et al. 2008; Ohm 2010; Parry 2011).

In cases of breaches of privacy, regardless of the cause, proportionality is important and failures need to be considered in context. Treating breaches with 'witch-hunts' and exorbitant fines may not have the desired effect. Rather than creating a positive environment for safe data sharing, we create a culture of fear and lockdown, with inadequate systems and individuals unwilling to take responsibility. Researchers have argued that this has resulted in a tense environment, in which it becomes:

> ... easy for the public, and regulators, to lose sight of how easily the increasingly broad body of restrictions limiting access to medical and public health data can

undermine efforts to better understand and improve public health.

(Wartenberg & Thompson 2010)

In medicine, we are learning that 'naming, shaming, and blaming' does not contribute to a safety culture, and this is a lesson that also needs to be learnt when it comes to data (Leslie 2014).

Our Future Selves

As the saying goes, our future self is the first recipient of shared data. Imagine trying to work with your data a year or two down the line – perhaps while writing up a thesis or perhaps while finally getting round to sorting out the revisions on a paper. You don't want to be dealing with a smattering of unlabelled disks, containing a bunch of old files in unrecognised formats, on a desk that belongs to a previous employer. If data is well described, organised, and in non-proprietary formats, it will be easier to sort through, share and reuse.

Often we are required to register with a project or ethics committee prior to collecting or accessing health data, so it makes sense to sketch out a data management plan at this point. There are resources on the web to help create the plan, such as the Digital Curation Centre's DMPonline[3] and the UK Data Archive's Data Management Checklist (UK Data Archive 2014). Best practice is developing rapidly, so a specialist such as an academic librarian or local information governance manager should be involved where possible.

If a project requires consent from participants, it is important to clearly set out any intentions for data sharing. Good

---

[3] https://dmponline.dcc.ac.uk/

communication is crucial and keeping people informed from the outset will help to establish trust. Where it is not possible to obtain consent, or where consent has not been obtained for retrospective data, a local ethics committee should be approached for advice (Hrynaszkiewicz et al. 2010). Approval by the committee may be given where data is anonymised, but care is needed to maintain privacy. The British Medical Association offers a toolkit outlining key factors to take into account when sharing data, and the UK Information Commissioner's Office offers an overview of approaches to anonymisation (British Medical Association 2014; Information Commissioner's Office 2012).

Data that is not properly described is unlikely to be reused, so good metadata is vital. At the simplest level, metadata can be a description of the important details of the data. Reviewing data papers, such as those published in *Open Health Data* and *Scientific Data*,[4] may help to identify useful information to include. More formal metadata standards are established according to discipline and should be adopted where appropriate. A directory of standards is maintained by the Digital Curation Centre (Digital Curation Centre 2014). In cases where data cannot be shared due to privacy issues, it should almost always be possible to share the descriptive metadata, making the data discoverable and potentially reusable.

In terms of data publication, there are an increasing number of options, such as creating new instances of web-accessible databases (for example, via DataVerse), depositing in an institutional repository, or sharing via data publishers such as Dryad and figshare (King & Crosas 2014).[5] Most importantly, the service

---

[4] *Open Health Data*: http://openhealthdata.metajnl.com/; *Scientific Data*: http://www.nature.com/sdata/
[5] Dryad: http://datadryad.org/; figshare: http://figshare.com/

should offer some reassurance that data will be sustained for the foreseeable future, and a unique identifier such as a digital object identifier (DOI) should be provided to enable accurate citation and tracking of reuse.

Anyone sharing data along these lines is leading the way, at the front of a community that is working towards better, collaborative science. With mechanisms for researchers to cite data, and funders increasingly recognising the importance of data sharing, a system that gives proper recognition to those who share data must now be on the horizon.

Acknowledgements

Thanks to Ross Mounce, Cameron Neylon, and Paul Wicks for helpful comments via social media, and to Samuel Moore and John Wilbanks for giving feedback on the content before publication.

References

Allen, K. (2012). Un-needed branded drugs "cost NHS millions." *Financial Times Data*. Retrieved August 06, 2014, from http://blogs.ft.com/ftdata/2012/12/06/un-needed-branded-drugs-cost-nhs-millions/.

British Medical Association. (2014). Confidentiality and disclosure of health information tool kit. Retrieved August 05, 2014, from http://bma.org.uk/practical-support-at-work/ethics/confidentiality-tool-kit.

Celi, L. (2014). The Outing of the Medical Profession: Data marathons to open clinical Research Gates to Frontline Service Providers. *London School of Economics Impact Blog*. Retrieved August 06, 2014, from http://blogs.lse.ac.uk/.

Digital Curation Centre. (2014a). DMPonline. Retrieved August 05, 2014, from https://dmponline.dcc.ac.uk/.

Digital Curation Centre. (2014b). List of Metadata Standards. Retrieved August 06, 2014, from http://www.dcc.ac.uk/resources/metadata-standards/list.

Dryad Data. (2014). Retrieved August 06, 2014, from http://datadryad.org/.

El Emam, K., Arbuckle, L., Koru, G., Eze, B., Gaudette, L., Neri, E., … Gluck, J. (2012). De-identification methods for open health data: the case of the Heritage Health Prize claims dataset. *Journal of Medical Internet Research*, *14*(1), e33. doi:10.2196/jmir.2001.

European Court of Human Rights. I v. Finland (2008).

figshare. (2014). Retrieved August 06, 2014, from http://figshare.com/.

Howard, A. (2012). The risks and rewards of a health data commons. *O'Reilly Radar*. Retrieved August 05, 2014, from http://radar.oreilly.com/2012/08/health-data-commons.html.

Hrynaszkiewicz, I., & Altman, D. G. (2009). Towards agreement on best practice for publishing raw clinical trial data. *Trials*, *10*, 17. doi:10.1186/1745-6215-10-17.

Hrynaszkiewicz, I., & Cockerill, M. J. (2012). Open by default: a proposed copyright license and waiver agreement for open access research and data in peer-reviewed journals. *BMC Research Notes*, *5*(1), 494. doi:10.1186/1756-0500-5-494.

Hrynaszkiewicz, I., Norton, M. L., Vickers, A. J., & Altman, D. G. (2010). Preparing raw clinical data for publication: guidance for journal editors, authors, and peer reviewers. *Trials*, *11*, 9. doi:10.1186/1745-6215-11-9.

Information Commissioner's Office. (2012a). *Anonymisation: Managing Data Protection Risk Code of Practice* (p. 108).

Information Commissioner's Office. (2012b). *Anonymisation: managing data protection risk code of practice*. ICO. Retrieved from http://ico.org.uk/for_organisations/data_protection/topic_guides/anonymisation.

Institute of Medicine. (2000). *To Err Is Human*. (Committee on Quality of Health Care in America, Ed.) (p. 312). Washington D.C.: National Academy Press.

Ioannidis, J. P. A. (2013). How many contemporary medical practices are worse than doing nothing or doing less? *Mayo Clinic Proceedings*, *88*(8), 779–81. doi:10.1016/j.mayocp.2013.05.010.

Ioannidis, J. P., Greenland, S., Hlatky, M. A., Khoury, M. J., Macleod, M. R., Moher, D., ... Tibshirani, R. (2014). Increasing value and reducing waste in research design, conduct, and analysis. *Lancet*, *383*(9912), 166–75. doi:10.1016/S0140-6736(13)62227-8.

King, G., & Crosas, M. (2014). The Dataverse Project. Retrieved August 06, 2014, from http://datascience.iq.harvard.edu/dataverse.

Korn, N., & Oppenheim, C. (2011). *Licensing Open Data: A Practical Guide (Version 2.0)*. Retrieved from http://discovery.ac.uk/files/pdf/Licensing_Open_Data_A_Practical_Guide.pdf.

Leslie, I. (2014, June). How mistakes can save lives: one man's mission to revolutionise the NHS. *New Statesman*. Retrieved from http://www.newstatesman.com/2014/05/how-mistakes-can-save-lives.

National Research Council. (2013). *U.S. Health in International Perspective: Shorter Lives, Poorer Health*. The National Academies Press. Retrieved from http://www.nap.edu/openbook.php?record_id=13497.

Neamatullah, I., Douglass, M. M., Lehman, L. H., Reisner, A., Villarroel, M., Long, W. J., ... Clifford, G. D. (2008). Automated de-identification of free-text medical records. *BMC Medical Informatics and Decision Making*, *8*, 32. doi:10.1186/1472-6947-8-32.

Ohm, P. (2010). Broken Promises of Privacy: Responding to the Surprising Failure of Anonymization. *UCLA Law Review*, *57*, 1701. Retrieved from http://papers.ssrn.com/abstract=1450006.

Open Knowledge. (2013). Open Definition. *Open Knowledge*. Retrieved August 04, 2014, from http://opendefinition.org/od/.

Open Knowledge. (2014a). Bad Data. Retrieved August 06, 2014, from http://okfnlabs.org/bad-data/.

Open Knowledge. (2014b). Open Data Index. Retrieved August 06, 2014, from https://index.okfn.org/about/#criteria.

Park, T., & VanRoekel, S. (2013). Introducing: Project Open Data. *The White House*. Retrieved August 04, 2014, from http://www.whitehouse.gov/blog/2013/05/16/introducing-project-open-data.

Parry, M. (2011). Harvard's Privacy Meltdown. *The Chronicle of Higher Education*. Retrieved August 05, 2014, from http://chronicle.com/article/Harvards-Privacy-Meltdown/128166/.

The Economist. (2013, October). Unreliable research: Trouble at the lab. *The Economist*. Retrieved from http://www.economist.com/node/21588057/print.

The Royal Society. (2012). *Science as an open enterprise: Final report*. London. Retrieved from https://royalsociety.org/policy/projects/science-public-enterprise/report/.

UK Data Archive. (2014). Data Management Checklist. Retrieved August 05, 2014, from http://www.data-archive.ac.uk/create-manage/planning-for-sharing/data-management-checklist.

UK Parliament. Data Protection Act 1998. Her Majesty's Stationery Office (1998). United Kingdom of Great Britain and Northern Ireland.

United States Congress. Health Insurance Portability and Accountability Act (1996). United States.

Wartenberg, D., & Thompson, W. D. (2010). Privacy versus public health: the impact of current confidentiality rules. *American Journal of Public Health*, *100*(3), 407–12. doi:10.2105/AJPH.2009.166249.

Wellcome Trust. (2014). *Establishing incentives and changing cultures to support data access*.

Zhao, M., Wang, P., Guan, Y., Cen, Z., Zhao, X., Christner, M., … Wang, J. (2011). Open-Source Genomic Analysis of Shiga-Toxin–Producing.

Open Research Data in Economics

Velichka Dimitrova

Open Economics Working Group, Open Knowledge
Foundation, London, United Kingdom

Just a few decades ago, particularly in the 1970s and 1980s, empirical work in economics lacked credibility: modifications to functional form, sample size or controls could change the findings and conclusions. Edward Leamer (1983) criticised the fragility of econometric results, saying that to draw inferences from data described in econometric texts, it was necessary "to make whimsical assumptions". For a long time nobody trusted the results of econometric papers.

Since then, better research designs, experiments or good quasi-experiments has lead to a credibility revolution in economics (Angrist and Pischke 2010) and "taking the "con" out

How to cite this book chapter:
Dimitrova, V. 2014. Open Research Data in Economics. In: Moore, S. A. (ed.) *Issues in Open Research Data*. Pp. 141–150. London: Ubiquity Press. DOI: http://dx.doi.org/10.5334/ban.i

of econometrics". Leamer's judgement of the empirical work of his time – that "hardly anyone takes anyone else's data analysis seriously" seems to be less justified today, largely due to quality research designs. Miguel et al. (2014) argue that these changes have been particularly pronounced in development economics with a large number of randomised trials in recent years.

Parallel to this trend, new opportunities of gathering and processing data has made some researchers enthusiastic about the opportunities to create novel research designs, to analyse large and granular datasets, allowing for better measurements of economic effects and outcomes, etc. (Einav and Levin 2013).

Data itself has become "the new oil" or "a new asset class" (Schwab et al. 2011). In many sub-disciplines of economics, the surging number of empirical papers has attested that greater availability of micro data has permitted "rigorous empirical analyses of questions that cannot be answered purely based on theory" (Raj and Finkelstein 2012).

The hype about all the opportunities which data creates often pays less need to the questions of access to data, reproducible research and transparency. Who if not economists understand the value generated by having open access to knowledge and data as well as the benefits of knowledge as a public good?[1].

Making economics research data and code available serves to enable scholarly enquiry and debate and to ensure that the results of economics research can be reproduced and verified. This is the

---

[1] Having the properties of non-rivalrousness and non-excludability, knowledge could be considered a public good or at least an "impure public good" as returns to some forms of knowledge can be appropriated to some extent (Stiglitz 1999).

rationale behind the Open Economics Principles[2], a Statement on the Openness of Data and Code – http://openeconomics.net/principles/. The purpose of the Principles is to provide some basic guidelines on why, how and when data in economic research should be open.

The first Principle is to have **"open data by default"** where *"data in its different stages and formats, program code, experimental instructions and metadata – all of the evidence used by economists to support underlying claims – should be open as per the Open Definition[3], free for anyone to use, reuse and redistribute"*. Having open data by default sets a gold standard for research in economics, where any researcher would have to abide by this principle where possible. Some empirical economists do provide access to their data and code on their websites and actively encourage their research to be replicated (where Joshua Angrist's data archive[4] is a leading example), yet there are still relatively few who do so.

Whilst many initiatives exist in the field of the natural sciences, social scientists and economists have been more hesitant about opening up data and code. Economists work with diverse and often sensitive data. Original empirical work depends on having unique datasets with individuals, households or firms as observation units. Such data may contain sensitive information or may be subject to confidentiality agreements. The researchers may also not own the data they work with.

---

[2] The Open Economics Principles were created by The Open Economics Working Group of the Open Knowledge Foundation. The statement was brought forward by an Advisory Panel (http://openeconomics.net/advisory-panel/) of economics professors, funders and practitioners with the support of the Alfred P. Sloan Foundation.

[3] http://opendefinition.org/

[4] http://economics.mit.edu/faculty/angrist/data1/data

Therefore, the second Principle recognises that *"there are often cases where for reasons **of privacy, national security and commercial confidentiality** the full data cannot be made openly available. In such cases researchers should share analysis under the least restrictive terms consistent with legal requirements, and abiding by the research ethics and guidelines of their community"*. Researchers would still be encouraged to open up non-sensitive data, summary data, metadata and code where applicable, as legal agreements may often allow for some degree of sharing.

Privacy and confidentiality are, however, not the only reasons for not opening up data and code. Access to quality and high-frequency data is often not free and requires significant investment of research resources. Gathering particular novel datasets requires a resource investment and researchers may not be willing to share data until they have exhausted all returns associated with their investment.

For that reason, the third Principle attempts to summarise the need to offer a reward associated with sharing as it deals with **reward structures and data citations** – *"recognizing the importance of data and code to the discipline, reward structures should be established in order to recognise these scholarly contributions with appropriate credit and citation in an acknowledgement that producing data and code with the documentation that make them reusable by others requires a significant commitment of time and resources"*.

The Principles also draw attention to the efforts of data curators which often are under-appreciated, but who have a major role in supporting researchers in gathering, documenting, storing and sharing research data.

Further rewards associated with the sharing of data and come may become more common in the future. Data citations are seen as a way to reward the efforts of researchers in producing data and making it easier for others to find and access datasets. If researchers begin to cite data the same way they cite articles and books, it would allow for tracking the data's impact, verifying and re-using it as well as acknowledging the contribution of the data producers[5].

Nevertheless, datasets in economics are most frequently related to one or more academic papers. The data and code serve to enable the verification of empirical results. Thus, the fourth Principle deals with the **data availability**: *"Investigators should share their data by the time of publication of initial results of analyses of the data, except in compelling circumstances. Data relevant to public policy should be shared as quickly and widely as possible. Funders, journals and their editorial boards should put in place and enforce data availability policies requiring data, code and any other relevant information to be made openly available as soon as possible and at latest upon publication."*

Recognising that data and code should be made available, economics journals have put in place data availability policies. The American Economic Review, which could be seen as setting the tone for the policy of other journals[6], requires the authors of accepted empirical papers to provide prior to publication all necessary data and computation necessary for replication and promises to make it available on the AER website[7]. Accordingly, the majority of the more recent AER articles have their datasets available online.

---

[5] See the DataCite project for details: https://www.datacite.org/
[6] The project EdaWaX evaluated the data availability of economics journals: http://openeconomics.net/resources/data-policies-of-economic-journals/
[7] The data availability policy of the American Economic Review: http://www.aeaweb.org/aer/data.php

In fact, the availability of raw data related to a paper is not a new issue. In what became to be regarded as the first referee report of an article submitted to Econometrica, Ragner Frisch commented on the work of Henry Schulz in October 1932:

"I would also suggest that you include a table giving the raw data you have used. ... I think the publishing of the raw data is very important in order to stimulate criticism and control" (Bjerkholt 2013).

Another emerging area is the pre-registration of economics and social science studies, especially where experiments are involved. For instance if the researchers are running a randomised controlled trial, they would have to state in their trial protocol what kind of outcomes they would like to observe. The more outcomes we look at, the more probable it is that there would be some indicator with a significant effect size. Stating *ex-ante* what the purpose of the trial is and what outcomes will be observed sets out a transparent research process.

The American Economic Association (AEA) launched in 2013 a registry for randomized controlled trials in economics (https://www.socialscienceregistry.org/) "to address the growing number of requests for registration by funders and peer reviewers, make access to results easier and more transparent, and help solve the problem of publication bias by providing a single place where all trials are registered in advance of their start"[8]. Pre-registration would help improve the quality of randomized experiments and tackle the selective presentation of results, the inadequate documentation of hypothesis testing and data mining.

---

[8] See short announcement at http://openeconomics.net/2013/07/04/the-aea-registry-for-randomized-controlled-trials/

Funders have also established data management and sharing plans where researchers are required to outline their approach to gathering, storing and disseminating their research data. However, many funders have to face the trade-off between giving more research funding and setting aside a pot for supporting the documentation of research. In line with these developments the U.S. government released a policy memorandum[9], promising specific funding for making federally-funded research freely available to the public, giving specific attention to digital data.

The Economic and Social Research Council in the UK also requires data management and data sharing plans from all grant applicants and recognises that data sharing and re-use are "becoming increasingly important"[10]. Funders of research are aware that having research and its underlying data out in the open has the potential of multiply the impact of the original project, thus making better use of the research resources.

The fifth Principle refers to the **openness of publicly funded data**: "*publicly funded research work that generates or uses data should ensure that the data is open, free to use, reuse and redistribute under an open license – and specifically, it should not be kept unavailable or sold under a proprietary license. Funding agencies and organizations disbursing public funds have a central role to play and should establish policies and mandates that support these principles, including appropriate costs for long-term data availability in the funding of research and the evaluation of such policies, and independent funding for systematic evaluation of open data policies and use.*"

---

[9] http://www.whitehouse.gov/sites/default/files/microsites/ostp/ostp_public_access_memo_2013.pdf

[10] ESRC Research Data Policy, September 2010 – http://www.esrc.ac.uk/_images/Research_Data_Policy_2010_tcm8-4595.pdf

As publicly funded research is done in the public interest, it should be also open for the public to access, as the greatest benefit would be realised when data and code are made open and publicly available. The analysis done by economists and social scientists is also often used to inform policy-making and serves as evidence for government interventions or de-regulation. Public engagement and trust are some of the underlying reasons for making economics research data and code openly available.

Economists like Reinhart and Rogoff[11] as well as Piketty[12] who have come under scrutiny with regard to their research methodology and data have had publish corrections or respond to criticisms. Where economic research results are adopted as recommendations in policy-making, it is essential that the methodology and data underlying these results can be reviewed and scrutinised. A lot of the economics evidence base may remain undiscovered or unused if not published in the proper way.

Therefore, the sixth Principle deals with **usability and discoverability**: *"as simply making data available may not be sufficient for reusing it, data publishers and repository managers should endeavour to also make the data usable and discoverable by others for example: documentation, the use of standard code lists, etc., all help make data more interoperable and reusable and submission of the data to standard registries and of common metadata enable greater discoverability"*.

Better systems and frameworks have emerged to encourage and enable the sharing of data and code. Projects like the Open Science Framework (https://osf.io/) provide platforms to researchers for

---
[11] http://www.ft.com/cms/s/0/433778c4-b7e8-11e2-9f1a-00144feabdc0.html#axzz3DJfB0D2P
[12] http://www.nytimes.com/2014/05/30/upshot/thomas-piketty-responds-to-criticism-of-his-data.html?_r=0&abt=0002&abg=0

storing and sharing their data throughout the research lifecycle, with the aim to increase productivity of academics and the efficiency of sharing. Web-hosting services with revision controls systems may be a model for collaboration projects also in the social sciences where researchers would be able to share their code and work more effectively.

Further tools exist for economic researchers to share their research data, e.g. projects like DataVerse at Harvard (http://thedata.org/) offer online repositories for research data. It is generally not the lack of available tools, which hinders openness of economic data.

There are many potential benefits for sharing data: it enhances the visibility and the impact of one's research: it allows for the scrutiny of research findings, promotes new uses of the data and avoids unnecessary costs for duplicate research. The revolution of credibility in econometrics needs to embrace open data in order to realise its full potential.

References

Angrist, Joshua D., and Jörn-Steffen Pischke. "The Credibility Revolution in Empirical Economics: How Better Research Design Is Taking the Con out of Econometrics." *Journal of Economic Perspectives* 24.2 (2010): 3–30.

Bjerkholt, Olav. *Promoting econometrics through Econometrica 1933–37*. No. 28/2013. Memorandum, Department of Economics, University of Oslo, (2013).

Chetty Raj and Finkelstein Amy "Program Report: The Changing Focus of Public Economics Research, 1980-2010" NBER Reporter (2012).

Einav, Liran, and Jonathan D. Levin. "The data revolution and economic analysis" National Bureau of Economic Research (2013).

Leamer, Edward E. "Let's Take the Con out of Econometrics." *The American Economic Review* 73.1 (1983): 31–43.

Miguel, E., et al. "Promoting Transparency in Social Science Research" *Science* (2014): 30–31.

Schwab, K., et al. "Personal Data: The Emergence of a New Asset Class." World Economic Forum Report. (2011).

Stiglitz, Joseph E. "Knowledge as a global public good." *Global public goods: International cooperation in the 21st century* 308 (1999): 308–25.

Open Data and Palaeontology

Ross Mounce

University of Bath, Bath, UK

Introduction

Palaeontology is the study of ancient life in all its forms: vertebrates, arthropods, plants and many other weird and wonderful types of organism. As an academic discipline, it suffers from a perception in some quarters that it is a less quantitative, less analytical, 'soft science'—a kind of Rutherfordian-view that the study of fossils is just 'stamp collecting'. Yet modern palaeontology is often highly computational, generating lots of data with which to test and form hypotheses. In the digital age, once published, if provided in the right format, data can be easily reused by further

How to cite this book chapter:
Mounce, R. 2014. Open Data and Palaeontology. In: Moore, S. A. (ed.) *Issues in Open Research Data*. Pp. 151–164. London: Ubiquity Press. DOI: http://dx.doi.org/10.5334/ban.j

studies to advance the sum of all human knowledge. This chapter examines the availability of palaeontology-related research data online and the reuse conditions under which it is made available.

Example Data Generating Studies in Modern Palaeontology

A typical study in systematic palaeontology may attempt to retrace the relationships between extinct life forms using an evolutionary tree (phylogeny). The source data in this instance may be a matrix of many thousands of observations of the morphology of fossil forms, codified into discrete states for analysis. These observations often come from comparative examination of specimens or, more likely, high-resolution photographs of these specimens that enable features to be examined side-by-side even if the physical specimens themselves are kept continents-apart in different museums.

Other palaeontological studies go one further and aim to create 'virtual fossils'—accurate three-dimensional interactive vizualisations of specimens to aid their interpretation, with the aid of tomographic methods. Methods such as X-ray imaging and magnetic resonance imaging (MRI) generate data non-destructively, so the original fossil is preserved undamaged. Both these types of palaeontological study represent just a small subset of the full range of palaeontological studies but what they have in common is that they heavily rely on imaging data; either photographs of specimens in the first instance, or the creation of three-dimensional image data. Much of palaeontology thus relies on the interpretation of morphology and thus image data, and the online sharing of image data is crucial to advancing palaeontological science.

Infrastructure Enabling Data Sharing in Palaeontology

There are many specialist sites specifically catering for or allowing palaeontological data, some of which incorporate helpful data management, collaboration and analysis tools that further incentivise use of their platform. I do not pretend to provide an exhaustive listing here—there are no doubt many more, the projects discussed herein reflect my own personal biases towards vertebrate palaeontology and systematic palaeontology. The main point of this selection is to highlight the variance in approach to data licencing that each of these projects has adopted. See 'From card catalogs to computers: Databases in Vertebrate Paleontology' for a review with a different focus (Uhen et al. 2013).

The Paleobiology Database
http://paleobiodb.org/

This project collates taxonomic and collection-based occurrence data for all fossil groups, of all geological ages. It is widely supported and contributed to by the palaeontological research community.

Towards the end of 2013 (Kishnor & Peters 2013), it set a great example by uniformly re-licencing all the data it contains under the Creative Commons Attribution (CC BY) 4.0 International License to ensure that it provides open, reusable data.

Their frequently asked questions (FAQs) (Alroy, adapted by Uhen 2013) suggest that for large (how large is left undefined) dataset analyses, data reusers should download an accompanying 'secondary bibliography' to provide evidence of data provenance for subsequent journal publication as a supplementary material file. Whilst this strategy certainly fulfils the legal requirements of the CC BY licence, such a request is extremely unlikely to provide counted citations, which help researchers demonstrate their academic impact. Most of the traditional bibliometric data indexers, e.g. Thompson Reuters Web of Knowledge and Google Scholar, only index the main paper for citations. Citations provided in supplementary files are typically ignored (Kueffer et al. 2011).

Ancient Human Occupation of Britain Database (AHOB)
http://www.ahobproject.org/database/

This project documents data on British and European Quaternary dig sites: geographical co-ordinates, photographs, stable-isotope data, faunal lists and more. It has received funding from three Leverhulme Trust programme grants.

Access is entirely restricted to project members-only for the life of the project. According to Uhen et al. (2013) the data '… will be made publicly available at the end of the project in 2013.' Yet in 2014 the database is still access-restricted, project member login-only. Licencing of the data contained in this database is unknown. Even if some of the data cannot be shared openly because it might be sensitive, it strikes me that at least some of the data, e.g. faunal lists and stable-isotope data, is clearly non-sensitive and therefore can without doubt be reasonably made publicly available.

MorphoBank
http://www.morphobank.org/

MorphoBank (O'Leary & Kaufman, 2011) is a website primarily used by researchers concerned with morphology-based phylogenetics or cladistics research (reconstructing evolutionary trees).

It has strong features that help researchers build, version control, annotate, manage and enable effective collaboration around their phylogenetic research data, as well as providing a web-space in which to make all that data publicly available after publication of the associated research paper. As of early April 2014, there are over 300 publicly accessible projects on MorphoBank as well as over 600 non-public projects in progress. The *Journal of Vertebrate Paleontology* should be congratulated as one of the first journals to publicly support the use of MorphoBank (Berta & Barrett 2011); as a result of this, there are more MorphoBank projects with data from *Journal of Vertebrate Paleontology*-published studies than any other journal.

Initially, data uploaded to MorphoBank is private, until researchers are ready to choose to make it public. When making their data public, MorphoBank allows researchers to choose from the full range of Creative Commons licences available. MorphoBank guides users towards choosing open licences on their FAQ but does not enforce their preference:

> MorphoBank would prefer for content providers to choose CC0 or CC BY reuse policies because they (and only they) are Open Data licenses. Please be aware that choosing an NC (non-commercial usage only) license may prevent your data submission from being used on open-content only websites such as Wikipedia.
> (MorphoBank 2014)

It is difficult to search media by licence, but I estimate (supporting data on figshare; Mounce 2014) that of the >27,000 publicly viewable images hosted on MorphoBank, less than half are made available under Open Knowledge Definition (OKD)-conformant open licences (see **Figure 1**). Over 77% of projects share less than 10 images, with most (modal) sharing only one image—MorphoBank forces users to upload at least one image.

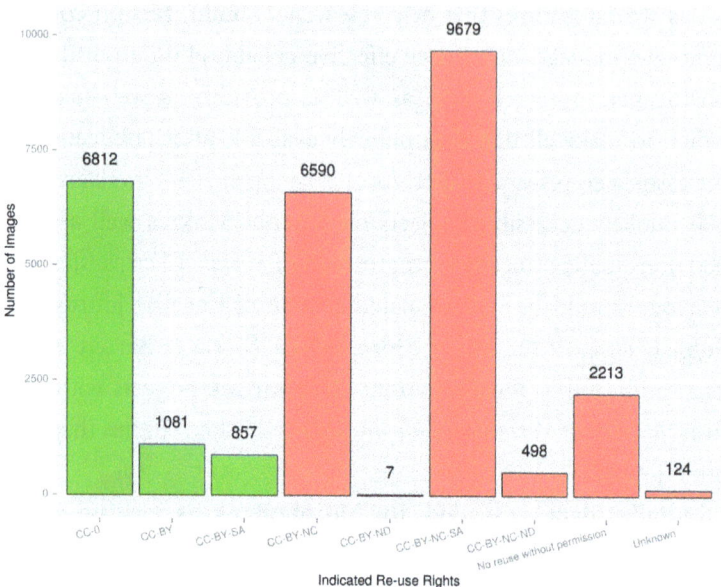

Figure 1: Images in MorphoBank by re-use rights. The three leftmost columns in green indicate OKD-conformant open licences. Figure generated in R (R Core Team, 2014) with the package ggplot2 (Wickham, 2009).

Morphbank
http://www.morphbank.net/

Not to be confused with its close namesake, Morphbank is an earlier project that specifically focuses on biological specimen image data sharing. As of early April 2014, this database makes publicly available over 372,000 images of biological specimens. By default, images are licenced under Creative Commons Attribution-NonCommercial-ShareAlike (CC BY-NC-SA; not an OKD-conformant open licence) but contributors may opt to change that for their uploads to a less restrictive Creative

Commons license, including even Public Domain Dedication. As with MorphoBank, it does not appear possible at this point to easily filter or search images by reuse licence so I am unable to determine the distribution of licences chosen by contributors to the site.

For some reason, however, few palaeontologists seem to have adopted the use of Morphbank to share their image data. Alberto Prieto-Marquez, a vertebrate palaeontologist, is one notable exception in that regard—he has made over 1700 images relating to his research available via this site.[1]

Dryad
http://datadryad.org/

Another more recent initiative to encourage data sharing that is open to palaeontologists is Dryad. All data submitted to Dryad is released to the public domain under the Creative Commons Zero waiver (CC0). The Paleontology Society journals (*Journal of Paleontology*, *Paleobiology*) were the first significant palaeontological adopters of Dryad, and now the palaeo-relevant journals *Palaeontology*, *ZooKeys* and *Zoological Systematics and Evolution* also make use of it to share supplementary, publication-associated data. The journal *Evolution* deserves special praise for being one of the first well-respected evolutionary biology journals to mandate data archiving for all its articles (Fairbairn 2011), something that many journals still just weakly 'encourage'. Key to the popularity of Dryad is probably its assignment of a digital object identifier (DOI) to each and every dataset contributed, which allows easier citation and tracking of the reuse of data. Of course, data does not actually need a DOI to be 'citable' but, for many, a DOI certainly

---

[1] User record available at http://www.morphbank.net/?id=78418

does seem to encourage formal citation. This may explain why some authors have even gone to the trouble of uploading datasets relating to long-ago published papers—something I would imagine they would not do if they saw no benefit to themselves in this service.

Figshare
http://figshare.com/

Figshare, similar to Dryad, is a 'generalist' data sharing website that is open to palaeontology but also contains data relating to a much wider array of subjects. Like Dryad, they also assign DOIs to datasets but they go one further in assigning each and every file within your data upload a separate DOI if you so wish. Unlike Dryad, figshare also allows the upload of data not related to publications, so it is ideal for uploading 'work-in-progress' data and data from projects that would otherwise be left in a file-drawer unfinished forever. I estimate at least 2000 research objects (figures, images, data, posters, manuscripts, code) relating to palaeontology have so far been made available at figshare. From a reuse rights perspective, figshare by default makes uploaded figures, media, posters, papers and filesets available under CC BY. Datasets are made available under CC0, and code under the MIT License. All these are OKD-conformant open licences.

Summary of Data Sharing Infrastructure for Palaeontology

As you can see from this small selection of palaeo-relevant databases, there is huge variance between them in terms of reuse rights. Some make nothing publicly available (e.g. AHOB), whilst many allow users to initially upload data privately and then make it publicly available at a later date (e.g. figshare, Dryad, MorphoBank, the Paleobiology Database). When data is made publicly available at these sites, some allow a wide choice of reuse rights options and content uploaders do typically make use of all of these options if options are provided (e.g. Morphbank and MorphoBank). Others such as figshare, Dryad and the Paleobiology Database have made a conscious and reasoned decision to not allow a choice of licences when making data available; all these three enforce OKD-conformant licenses—either CC BY or CC0.

Interestingly, prior to the late 2013 licencing change by the Paleobiology Database committee, PaleoDB (as it was then known) used to allow data contributors to upload data under a variety of different Creative Commons licences. Many contributors chose different licences, and some of these licences were incompatible with each other! This along with many other reasons (given in Hagedorn et al. 2011; Klimpel 2012) is why PaleobioDB opted to adopt CC BY only.

Is licence choice really a good thing?

Having content available in a variety of different licences in projects such as at Morphbank and MorphoBank creates a lot of additional complexity for bulk reusers of content. Having to accommodate this variability is hard, especially if some of those different terms and conditions are incompatible with each other. Databases such as Dryad that use CC0 impose no legal restraint on data reuse, and instead trust academic cultural norms to ensure that data is cited appropriately if reused. I am confident that in science we do not need to resort to copyright-led enforcement of citation, and that academic cultural norms and the self-policing nature of academia are enough to ensure citation from data reuse. As testament to this, I know of no instances in which data made available at Dryad or figshare has been reused without appropriate citation.

Another troubling aspect is the seemingly widespread adoption of the 'non-commercial' (-NC) Creative Commons licences where they are allowed. I suspect this is based upon misunderstanding of the type of reuse(s) that these licences prevent. Many assume that non-commercial licences only prevent for-profit businesses from reusing content for profit. But non-commercial is about commerce, not profit, and that is an important difference. In my experience, few realise that these non-commercial licences are far more restrictive: -NC content cannot be reused in most educational settings in schools or universities, likewise -NC content cannot be uploaded to Wikimedia for use on Wikimedia projects like Wikipedia (Klimpel 2012). Indeed, a recent ruling in Germany shows that -NC content is only 'safe' for personal use (Haddouti 2014): any other use, even by a non-profit organisation, may get the content reuser sued many years later. Myself and many others would not want to expose ourselves to this risk and thus -NC licenced content is unusable for us.

The Role of Journals in Encouraging Data Sharing

In my opinion I see journal policy as key to encouraging and enforcing data sharing. There are the beginnings of a trend to be observed in which the better journals mandate the archiving of all publication-related supporting data to encourage its examination and reuse (Fairbairn 2011). This is in both the authors' and journals' interests because sharing data is known to be associated with an increased citation rate (Piwowar, Day & Fridsma 2007; Piwowar & Vision 2013), as well as being cost-effective (Piwowar, Vision & Whitlock 2011). I would like to think these advantages alone would facilitate spontaneous data sharing, but I do not see that happening in the palaeontological community, so research-funder and journal policies are still needed to encourage and enforce data sharing.

The journals *Evolution*, *Journal of Paleontology*, *Paleobiology* and *ZooKeys* clearly mandate that all data should be shared. Then there are a lot of journals like the *Journal of Vertebrate Paleontology* (Berta & Barrett 2011) that merely encourage full data archiving. Even within the same society there is policy variance: of the Linnean Society journals, the *Biological Journal of the Linnean Society* requires Dryad data archiving, whilst the *Zoological Journal of the Linnean Society* does not mandate data archiving, anywhere. I have had to contact the editor of the *Zoological Journal of the Linnean Society* many times with regards to data issues in that journal. It would help my research, and presumably many others, if *Zoological Journal of the Linnean Society* took a stronger approach with regards to its data sharing guidelines.

Conclusions

Palaeontological data and its availability in the digital era is an interesting subject with many ongoing developments. For many types of data that would concern palaeontologists, there are no unsolved technical barriers in the way of sharing data openly anymore; the only barrier is social adoption, willingness to share. For phylogenetic data there are well-established data standards such as Nexus and 'hennig' with which to exchange data in small plain text files, as well as specialist databases for it, e.g. MorphoBank. This phylogenetic data is increasingly being uploaded online. But for images and photographs the trend is different. Despite a much wider selection of databases available, I detect a certain reluctance from palaeontologists to upload their specimen research photographs in their entirety.

Palaeontology, and indeed all morphology-based biological research, is utterly dependent upon the interpretation of specimen morphology, so it is vital that photographic imagery of these specimens and their attributes are made available for all to see and use (Ramírez et al. 2007; Cranston et al 2014). Until full, high-resolution images of specimens are abundantly and openly available online, systematic palaeontology will continue to be an expensive endeavour, often requiring researchers to travel to museums all across the world to view and take photos of specimens they need for their comparative research. Thus, even despite the Internet, much of palaeontological research still operates in a kind of pre-Gutenberg manner akin to the age where scholars had to travel to each of the best libraries in the world to read books of which there were no copies anywhere else. The Internet has revolutionised the dissemination of written works, enabling their free and easy copying. But, for palaeontological specimens and research-quality images of them, the digital revolution has really yet to begin. For three-dimensional imaging, the many hundreds of gigabytes of raw tomographic data required for each specimen may seem to be a valid barrier for not sharing them openly online. However, I see no such good excuse as to why there are not more openly available high-resolution photographic images of palaeontological research specimens. The infrastructure is certainly in place and cost-efficient, if not 'free', for researchers—it just needs to be used!

References

Alroy, J adapted by Uhen, M D 2013 *Frequently Asked Questions* (Paleobiodb.org). Available at http://paleobiodb.org/#/faq/citations [Last accessed 14 August 2014].

Berta, A and Barrett, P M 2011 Editorial. *Journal of Vertebrate Paleontology*, 31(1): 1. DOI: http://dx.doi.org/10.1080/02724634.2011.546742.

Cranston, K, Harmon, L J, O'Leary, M A and Lisle C 2014 Best Practices for Data Sharing in Phylogenetic Research. *PLOS Currents Tree of Life*. 2014 Jun 19. Edition 1. doi:http://dx.doi.org/10.1371/currents.tol.bf01eff4a6b60ca4825c69293dc59645.

Fairbairn, D J 2011 The advent of mandatory data archiving. *Evolution*, 65: 1–2. DOI: http://dx.doi.org/10.1111/j.1558-5646.2010.01182.x.

Haddouti, C S 2014 *Verstoß Gegen CC-Lizenz: Deutschlandradio Muss Zahlen*. Available at http://www.heise.de/newsticker/meldung/Verstoss-gegen-CC-Lizenz-Deutschlandradio-muss-zahlen-2151308.html [Last accessed 14 August 2014].

Hagedorn, G, Mietchen, D, Morris, R, Agosti, D, Penev, L, Berendsohn, W and Hobern, D 2011 Creative Commons licenses and the non-commercial condition: implications for the re-use of biodiversity information. *ZooKeys*, 150: 127-149. DOI: http://dx.doi.org/10.3897/zookeys.150.2189.

Kishnor, P and Peters, S 2013 *Paleobiology Database Now CC BY*. Available at http://creativecommons.org/weblog/entry/41216 [Last accessed 13 August 2014].

Klimpel, P 2013 *Consequences, Risks, and Side-Effects of the License Module Non-Commercial – NC 1-22*. Available at http://openglam.org/files/2013/01/iRights_CC-NC_Guide_English.pdf.

Kueffer, C, Niinemets, Ã, Drenovsky, R E, Kattge, J, Milberg, P, Poorter, H, Reich, P B, Werner, C, Westoby, M and Wright, I J 2011 Fame, glory and neglect in meta-analyses. *Trends in Ecology & Evolution*, 26: 493-494. DOI: http://dx.doi.org/10.1016/j.tree.2011.07.007.

Mounce, R 2014 MorphoBank image content analysis. *Figshare*. DOI: http://dx.doi.org/10.6084/m9.figshare.994172.

O'Leary, M A and Kaufman, S 2011 MorphoBank: phylophenomics in the "cloud". *Cladistics* 27: 529-537. DOI: http://dx.doi.org/10.1111/j.1096-0031.2011.00355.x.

Piwowar, H A, Day, R S and Fridsma, D B 2007 Sharing detailed research data is associated with increased citation rate. *PLOS One*, 2: e308+. DOI: http://dx.doi.org/10.1371/journal.pone.0000308.

Piwowar, H A, Vision, T J and Whitlock, M C 2011 Data archiving is a good investment. *Nature*, 473: 285. DOI: http://dx.doi.org/10.1038/473285a.

Piwowar, H and Vision, T J 2013 Data reuse and the open data citation advantage. *PeerJ*, 1: e175. DOI: http://dx.doi.org/10.7717/peerj.175.

R Core Team 2014 R: A language and environment for statistical computing. R Foundation for Statistical Computing, Vienna, Austria. URL http://www.R-project.org/.

Ramírez, M J, Coddington, J A, Maddison, W P, Midford, P E, Prendini, L, Miller, J, Griswold, C E, Hormiga, G, Sierwald, P, Scharff, N, Benjamin, S P and Wheeler, W C 2007 Linking of digital images to phylogenetic data matrices using a morphological ontology. *Systematic Biology*, 56: 283-294. DOI: http://dx.doi.org/10.1080/10635150701313848.

The MorphoBank Project 2012 *FAQ*. Available at http://www.morphobank.org/index.php/FAQ/Index [Last accessed 13 August 2014].

Uhen, M D, Barnosky, A D, Bills, B, Blois, J, Carrano, M T, Carrasco, M A, Erickson, G M, Eronen, J T, Fortelius, M, Graham, R W, Grimm, E C, O'Leary, M A, Mast, A, Piel, W H, Polly, P D and Säilä, L K 2013 From card catalogs to computers: databases in vertebrate paleontology. *Journal of Vertebrate Paleontology*, 33: 13-28. DOI: http://dx.doi.org/10.1080/02724634.2012.716114.

Wickham, H 2009 ggplot2: elegant graphics for data analysis. Springer New York.

www.ingramcontent.com/pod-product-compliance
Lightning Source LLC
Chambersburg PA
CBHW062107080426
42734CB00012B/2783